KINU
Korea Institute for National Unification

Toward Greater Transparency in Non-Nuclear Policy: A Case of South Korea

Cheon Seongwhun

The Korea Institute for National Unification (KINU) is a non-profit government research organization commissioned to study issues regarding peace settlement on the Korean Peninsula and the unification of the two Koreas. It is contributing to the reconciliation and cooperation of the two Koreas as well as their unification through basic research on related affairs, the development of a policy on national unification, and the formation of a national consensus.

ISBN No.: 89-8479-296-9-93340   ₩ 7,000

Copyright 2005 KINU

Published 2005 by KINU

535-353 Suyu 6-dong, Gangbuk-gu, Seoul 142-887, South Korea
www.kinu.or.kr
To order KINU documents or to obtain additional information,
contact Distribution Services: Telephone 82-2-901-2559/2520,
or the Government Publication Sales Center 82-2-734-6818.

Toward Greater Transparency in Non-Nuclear Policy: A Case of South Korea / Cheon Seongwhun – Seoul: the Korea Institute for National Unification, 2005
p. ; cm. – (Studies series ; 05-01)

ISBN 89-8479-296-9-93340

349.9-KDC4
327.1747-DDC21   CIP2005002139

Toward Greater Transparency in Non-Nuclear Policy:
A Case of South Korea

About the Author

Cheon Seongwhun is a Senior Research Fellow at the Korea Institute for National Unification (KINU), Seoul, South Korea. He is the author of numerous books and reports including *Non-Nuclear Policy of the Unified Korea: Looking Beyond and Being in the Process of Unification* (Seoul: KINU, 2002), *U.S. National Missile Defense and South Korea's National Security* (Seoul: KINU, 2001), and *Cooperatively Enhancing Military Transparency on the Korean Peninsula: A Comprehensive Approach* (Albuquerque, NM: Cooperative Monitoring Center, SNL, 1999). He is the recipient of Commendation of President of the Republic of Korea in 2003, and awards for excellent research from the Korea Research Council for Humanities & Social Sciences, Office of the Prime Minister in 2001, 2002 and 2003.

* * *

Toward Greater Transparency in Non-Nuclear Policy: A Case of South Korea

Contents

Korea Institute for National Unification

Contents

Contents

Appendixes

References

Chapter One

The ROK and Nuclear Non-Proliferation in 2004

In August 2004, it was revealed that the Republic of Korea (ROK) failed to report its nuclear activities on several occasions from the early 1980s to the International Atomic Energy Agency (IAEA)—dubbed as "the 2004 incident" in this report. As the North Korea nuclear crisis intensified and the 7th Review Conference of the Non-Proliferation Treaty (NPT) was scheduled ahead in May 2005, the incident drew great attention from the international non-proliferation community. Although the quantities of nuclear material involved and the scales of nuclear activities were not significant, the nature of the activities and the failures of the ROK nuclear community to report timely to the Agency were regarded as a serious matter triggering concerns about possible negative repercussions on the international non-proliferation norms.

Facing nuclear-armed North Korea and being surrounded by nuclear or nuclear-capable big powers, a suspicion of the international society has been laid on South Korea for its inherent desire to develop nuclear weapons to counter against external nuclear threats. Although South Korea has upheld a strong commitment to nuclear non-proliferation as a backbone of its foreign and security policy since 1991, the 2004 incident demonstrated that South Korea's non-nuclear policy has not been effective enough to eliminate the international suspicion.[1] In the aftermath of the 2004 incident, a consensus was made both within and without South Korea that Seoul must exert more efforts to enhance transparency of its non-proliferation policy. For instance, the ROK government proclaimed, on September 18, 2004, the four principles of the peaceful uses of nuclear energy, one of which was to firmly maintain nuclear transparency and to strengthen international cooperation. In the international domain, the IAEA Director General requested South Korea to make every effort to provide further information about its nuclear activities[2] and the IAEA Board encouraged the South to continue its active cooperation with the Agency, pursuant to its safeguards agreement and additional protocol.[3]

Having the seriousness of the 2004 incident in mind, this study will conduct academic analyses of enhancing transparency in a key national policy area whose importance has been highlighted in the post Cold War era and examine practical implications for South Korea's non-nuclear policy. This study consists of five major parts. In chapter one, the 2004 incident is briefly summarized and lessons and reactions are discussed. Chapter two

---

[1] "Non-nuclear policy" means a comprehensive national policy to research, develop, and use nuclear energy only for the peaceful purposes and not to develop or possess nuclear weapons. Non-nuclear policy is defined to encompass nuclear non-proliferation policy.

[2] Article 40, *Implementation of the NPT safeguards agreement in the Republic of Korea*, GOV/2004/84, November 26, 2004.

[3] *IAEA Board of Governors Chairman's Conclusion on Implementation of the NPT Safeguards Agreement in the Republic of Korea*, November 26, 2004.

examines transparency in theoretical and practical perspectives: presenting various definitions of transparency, investigating functions of transparency, and explaining global phenomena of increasing policy transparency. In chapter three, the role of transparency is studied in the context of cooperative security framework. The importance of cooperative security has loomed large since the end of the Cold War. Chapter four suggests the main reasons why the ROK government needs to sustain its non-nuclear policy and further enhance its transparency. Finally, in chapter five, propositions are made to help the ROK government and nuclear energy community to formulate constructive and positive positions on nuclear transparency in line with international non-proliferation norms and standards. Major documents referred to in this study are gathered as appendixes at the end of the report.

Summary of the 2004 Incident

The ROK signed the NPT on July 1, 1968 and ratified it in April 1975. As a member of the Non-Proliferation Treaty, the Republic of Korea has adhered to the full-scope safeguards agreements. The Agreement between the ROK and the IAEA of safeguards in relation to the NPT entered into force on November 14, 1975.[4] As the nuclear proliferation concerns ride high as was manifested by Iraq and North Korea in the early 1990s, the IAEA has set to strengthen its safeguards system, leading to the additional protocol to the safeguards agreement in May 1997. The ROK signed the additional protocol on June 21, 1999, and ratified it on February 19, 2004. According to the additional protocol, the ROK has disclosed new information regarding its nuclear research and

---

[4] The official document of the agreement is *INFCIRC/236*.

development activities, triggering the 2004 incident.

On August 23, 2004, the South Korean government informed the IAEA that it had discovered, in June 2004 that laboratory scale experiments using the atomic vapour laser isotope separation (AVLIS) method for *enriching uranium* had been carried out in 2000 by a few scientists at the Korea Atomic Energy Research Institute (KAERI). The ROK government explained that the experiments had been conducted as a part of a broader scientific experiment to apply AVLIS techniques to non-nuclear materials like gadolinium and thallium. It also made it clear that only about 200mg of enriched uranium was produced and the installation and equipment were dismantled after the experiment was terminated.

Following the South Korean submission of the initial report according to the additional protocol and disclosure of the uranium enrichment, the IAEA began a series of verification missions in South Korea. During the inspections, the ROK further revealed that a few scientists had conducted *uranium conversion* activities in the 1980s that involved the production of about 154kg of natural uranium metal, a small amount of which was used in the AVLIS experiments in 2000.

In response to the IAEA enquiry, the ROK also disclosed that in the early 1980s, laboratory scale experiments had been performed at the TRIGA Mark III research reactor at the KAERI branch in Seoul. The purpose was to irradiate 2.5kg of depleted uranium and to study the *separation of uranium and plutonium*. On November 5, 2004, South Korea stated that 0.7g of plutonium was produced in the mini-assembly facility at KAERI.

At the Board of Governors meeting on September 13, 2004, South Korea's implementation of safeguards agreement was one of major issues. The Director General summarized the ongoing development relating to South Korea and concluded "it is a matter of serious concern that the conversion and enrichment of uranium and the separation of plutonium were not reported to the Agency as required by the ROK safeguards agreement."[5] During a later Q&A session, Dr. ElBaradei further elaborated that:[6]

> First of all we need to understand the nature and scope of the activities that took place in the Republic of Korea before we discuss what sort of action the Board needs to take. I think that the Board, at this stage, will simply ask me to continue to investigate the initial report we have received. And it will take us time, I would hope we can finish by November, but if not, then we will continue. Again, it depends on what we see; it depends of the level of co-operation we get from South Korea. But, as I said, so far, I am getting good transparency and good co-operation from Korea and I'll hope we should get a comprehensive report and get to the bottom of this issue by November.

In a response to an inquiry by the IAEA, the ROK provided additional information about a chemical enrichment experiment on October 21, 2004. It said that an experiment had been conducted during the period from 1979 through 1981 to assess a chemical exchange process to confirm the feasibility of producing 3% U-235. The Agency is in the process of assessing the ROK declaration of this experiment.

As a result of a series of inspections occurred in the later half of

---

[5] *Introductory Statement of the Board of Governors by IAEA Director General Dr. Mohamed ElBaradei*, September 13, 2004.
[6] *Transcript of the Director General's Press Statement on IAEA Inspection in Iran, Libya & the Republic of Korea*, September 13, 2004.

2004, the IAEA concluded that on a number of occasions, starting in 1982 and continuing until 2000, South Korea performed experiments and activities involving uranium conversion, uranium enrichment and plutonium separation and failed to report them to the IAEA in accordance with its obligations under its safeguard agreement. [7] In specific, the Agency listed the following four failures:

- Failure to report nuclear material used in evaporation, spectroscopy and enrichment experiments (AVLIS and chemical exchange) and the associated products;
- Failure to report the production, storage and use of all natural uranium metal and associated process loss of nuclear material, and the production and transfer of resulting waste;
- Failure to report the dissolution of an irradiated mini-assembly and the resulting uranium-plutonium solution, including the production and transfer of waste;
- Failure to report initial design information for the enrichment facilities and updated design information for the facilities involved in the plutonium separation experiment and the conversion to natural uranium and depleted uranium metal.

The Board of Governors discussed the ROK incident at the meeting in November 2004. After indept and extensive discussions, they adopted a conclusion spelling out the following points.[8]

---

[7] Article 38, *Implementation of the NPT safeguards agreement in the Republic of Korea*, GOV/2004/84, November 26, 2004.

[8] *IAEA Board of Governors Chairman's Conclusion on Implementation of the NPT Safeguards Agreement in the Republic of Korea*, November 26, 2004.

- The Board shared the Director General's view that given the nature of the nuclear activities described in his report, the failure of the Republic of Korea to report these activities in accordance with its safeguards agreements is of serious concern.
- The Board noted that the quantities of nuclear material involved have not been significant, and that to date there is no indication that the undeclared experiments have continued.
- The Board welcomed the corrective actions taken by the Republic of Korea, and the active cooperation it has provided to the Agency. The Board encouraged the Republic of Korea to continue its active cooperation with the Agency, pursuant to its safeguards agreement and additional protocol.
- The Board requested that the Director General report as appropriate.

Now the IAEA remains in the process of clearing residual uncertainties and the ROK is in full cooperation with the Agency. At the Board of Governors meeting in June 2005, nothing particular was mentioned about the 2004 incident, indicating international concerns had been allayed to a great extent.

Lessons and Reactions

Observing the South Korean incident, one study articulated a concern about possible negative influences on the international nuclear non-proliferation regimes:[9]

---

[9] Mark Gorwitz, *The South Korean Laser Isotope Separation Experience* (Washington, D.C.: Institute for Science and International Security, 2004), p. 1.

This transfer of knowledge represents a proliferation concern. Even though the uranium enrichment experiments were done on a laboratory level, once the physics have been mastered, scale-up efforts are seen as a matter of engineering. Armed with a competent core of engineers and scientists, the task of engineering a large-scale laser enrichment program is not beyond the current capabilities of South Korea.

It is noted that irresponsible and emotional coverage by some major foreign media reports contributed to exacerbating the 2004 incident and embarrassing the ROK government. Right after a portion of the ROK report to the IAEA that should have been confidential was leaked by the Nelson Report in the United States, the ROK Ministry of Science and Technolocy (MOST), on September 2, 2004 disclosed a main piece of the report in order to assuage rising international suspicions. The information was that a minuscule amount of enriched uranium—0.2 grams—was produced by one-off experiment test at the KAERI in 2000. It is astonishing that a representative American newspaper raised serious suspicions that this experiment was a part of South Korea's nuclear weapon development program and even called the scientists who performed the experiment "rogue scientists."[10] When prominent Japanese politicians—including former Prime Minister Yashiro Nakasone—occasionally have spoken for having nuclear weapons and a significant discrepancy was revealed in 2003 that 206 kilograms of plutonium was not accountable from Japan's reprocessing activities, it is wondered whether this newspaper would have labeled Japanese politicians and scientists with the adjective "rogue."

---

[10] David Sanger and William Broad, "South Koreans say secret work refined uranium," *New York Times*, September 3, 2004; James Brooke, "South Koreans repeat: we have no atom bomb program," *New York Times*, September 4, 2004.

The outbreak and subsequent development of the 2004 incident raised a doubt about whether the IAEA did well in assuring its employees to comply with their work ethics. According to the IAEA regulations, the employees are obliged to abide by the confidentiality of all the information received during their work and not to reveal any of them outside. Specifically, the IAEA Statute puts emphasis on integrity as a key criterion of recruiting its employees:[11]

> The paramount consideration in the recruitment and employment of the staff and in the determination of the conditions of service shall be to secure employees of the highest standards of efficiency, technical competence, and *integrity* [emphasis added].

The Statute also stipulates the following service regulations for the Agency staff:[12]

> In the performance of their duties, the Director General and the staff shall not seek or receive instructions from any source external to the Agency. They shall refrain from any action which might reflect on their position as officials of the Agency; subject to their responsibilities to the Agency, they shall not disclose any industrial secret or other confidential information coming to their knowledge by reason of their official duties for the Agency. Each member undertakes to respect the international character of the responsibilities of the Director General and the staff and shall not seek to influence them in the discharge of their duties.

Major news reports, however, frequently quoted sources indicating strong connections with the IAEA or its employees and revealed the details about the supposedly confidential ROK report

---

[11] Section D, Article VII—Staff, *Statute of the International Atomic Energy Agency*, October 26, 1956.
[12] Section F, Article VII—Staff, *Statute of the International Atomic Energy Agency*, October 26, 1956.

to the IAEA. For example, Washington Post quoted "IAEA sources" and "one diplomat familiar with the IAEA's work" to describe six violations identified by the IAEA.[13] An article published by the Arms Control Association quoted "a diplomatic source in Vienna close to the IAEA" to confirm a Washington Post article on September 12, 2004 that South Korea had enriched uranium to 77%.[14] The Washington Post article referred to diplomats and IAEA reports as sources of its reporting.[15] This problem of loose control of the IAEA employees needs be discussed in future IAEA gatherings as appropriate. And some measures against repeating a similar problem should be implemented for the benefit of strengthening the viability and sustainability of the international non-proliferation regimes as well as of maintaining fairness and objectivity of the IAEA.

In addition, the Japanese emotional reaction was not helpful for solving the 2004 incident in a more cooperative and calm environment. For example, only a few days after the incident began to be leaked into the public, the Japanese Minister of Government Hosoda Yukihiro defined it as an inappropriate matter that the ROK did not accept sufficient inspections.[16] He further urged the IAEA to carry out stringent inspections to the ROK, which could be interpreted as a political pressure to the Agency. The Asahi Shimbun also raised a suspicion that the ROK pursued a secret nuclear weapon development program in early

---

[13] Anthony Faiola and Dafna Linzer, "S. Korea admits extracting plutonium," *Washington Post*, September 10, 2004, p. A01.

[14] Paul Kerr, "IAEA continues investigation into South Korean nuclear activities," September 17, 2004, http://www.armscontrol.org.

[15] Dafna Linzer, "S. Korea nuclear project derailed," *Washington Post*, September 12, 2004, p. A24.

[16] *Donga Ilbo*, September 11, 2004.

1980s.[17] Later, the ROK expressed its regret at Japan's raising groundless suspicion of South Korea's nuclear activities. Seoul dispatched a high-level foreign ministry official to meet with Mr. Hosoda and explain its non-nuclear policy.[18]

All in all, the 2004 incident, in no way, should or could be interpreted as representing a desire by the ROK government to pursue a nuclear weapon development program.[19] The IAEA acknowledged that the quantities of nuclear material involved in the experiments have not been significant and that there was no indication of further undeclared experiments.[20] While participating in the Pugwash conference held in Seoul in October 2004, Dr. ElBaradei remarked that the South Korean case could not be comparable to the North Korean nuclear weapon development program.[21]

Nevertheless, the incident has produced several important implications.[22] First, an obvious conclusion is that the South Korean nuclear establishment is poorly regulated. The South Korean case is starkly compared with the Japanese example that Japan promptly reported to the Agency in 2003 the 206kg amount of plutonium—the discrepancy occurred in its reprocessing activities for a few decades. Given the mobile nature of nuclear-capable scientists and technicians, tightening up sloppy bureaucratic procedures and loose controls over nuclear assests—

---

[17] *Asahi Shimbun*, September 9, 2004.

[18] *Donga Ilbo*, October 11, 2004.

[19] Jungmin Kang, *et al.*, "South Korea's nuclear surprise," *Bulletin of the Atomic Scientists*, January/February 2005, pp. 40-49.

[20] Article 41, *Implementation of the NPT safeguards agreement in the Republic of Korea*, GOV/2004/84, November 26, 2004.

[21] *Donga Ilbo*, October 7, 2004.

[22] Jungmin Kang, *et al.*, "South Korea's nuclear surprise," pp. 40-49.

researchers, fissile materials, and technologies—is strongly needed. Second, the incident stirred already troubled waters in Northeast Asia due to North Korea's nuclear development. The South Korean case could let Japan and Taiwan rethink about the future of their non-nuclear status. Third, it was demonstrated that states will eventually pay a price if they allow nuclear research establishments to conduct activities without stringent education of the importance of adhering to non-proliferation norms and rules. Fourth, the incident manifests a proliferation impulse that arises out of discriminatory treatment. For example, the United States allows Japan to enjoy a full spectrum of nuclear activities while limiting South Korea's research activities. That is, there exists a disparity in the application of supposedly universal norms of non-proliferation. At the moment, there is no right cure to remedy this symptom of discrimination.

In the midst of rising international suspicions about past reporting failures and facing North Korean allegation of South Korea's secret nuclear weapon development,[23] the ROK government declared a new non-nuclear policy on September 18, 2004. Dubbed as "the four principles of peaceful uses of nuclear energy," this policy inherits spirits and commitments expressed in previous non-nuclear policies. The first was President Roh Tae Woo's declaration of denuclearization and settling peace on the Korean peninsula on November 8, 1991 and the second was the

---

[23] For example, the deputy at the DPRK mission at the United Nations Han Song Ryul criticized that the United States exercised a double standard favoring South Korea against North Korea and said that the South's uranium experiment was perceived as a threat to the North. *Yonhap News*, September 9, 2004. In his speech at the United Nations General Assembly on October 1, 2004, the DPRK Vice Foreign Minister Choe Su Hon argued that everything should be cleared about South Korea's secret nuclear weapon development program, which should be a condition for North Korean participation in the six-party talks. *Rodong Shinmun*, October 1, 2004.

Joint Declaration on Denuclearization of the Korean peninsula signed by North and South Korea on December 31, 1991. The new non-nuclear policy is a result of the 300th National Secuirty Council meeting and contains the following four major principles:

- Reconfirm the ROK government position that it has no intention to develop or possess nuclear weapons.
- Express the ROK government determination that it will firmly hold on to the principle of nuclear transparency and reinforce international cooperation, including full cooperation with the IAEA inspection.
- Reconfirm the ROK government intention that it will sincerely abide by international non-proliferation norms and rules such as the Non-Prolferation Treaty and the Denuclearization Declaration.
- Declare the ROK government desire to expand the peaceful uses of nuclear energy, based on international confidence obtained by the efforts according to the above three principles.

A specific policy measure resulted from the new non-nuclear policy. The National Nuclear Management and Control Agency (NNCA) was established in October 2004. The Ministry of Science and Technology decided to substitute the NNCA for the Technology Center for Nuclear Control (TCNC) that had been affiliated with the KAERI in order to strengthen national safeguards system and to enhance nuclear transparency. The NNCA was deliberately discharged from the KAERI so as to eliminate any question of its independence and neutrality as was raised to the TCNC. The NNCA is mandated to carry out missions such as technically supporting the MOST in safeguards, physical protection and export control. The NNCA consists of 6 major departments whose roles are as follows:[24]

- Planning & Policy: to develop the ROK nuclear policy on safeguards, physical protection, export control, and to develop external relations with the IAEA and other foreign countries.
- Nuclear Security and Protection: to review the regulations for physical protection of nuclear facilities, to construct and implement national physical protection regime, to develop methodology for threat assessment and regulatory technology, and to develop measures against radiation terrorism.
- Safeguards Implementation: to review regulations for accounting and control of nuclear materials, to implement inspections and evaluate inspection results, and to provide support for the IAEA inspection in the ROK.
- Verification Technology Development: to develop nuclear material non-destructive analysis methods, to conduct research on physical protection technology, to improve inspection equipment, and to construct remote monitoring systems.
- Information Management: to manage and analyze information on nuclear control systematically, including compliance with the IAEA additional protocol, and to analyze and control information on export and import.
- Public Relation and Education: to develop and implement compulsory education plans regarding nuclear control for South Korean nuclear scientists and workers, and to increase public awareness of the importance of nuclear transparency.

---

[24] *NNCA Newsletter*, January/February 2005, Daejeon, South Korea, http://www.nnca.re.kr.

Chapter Two

Transparency: Concept and Phenomena

Transparency began to draw the attention to academic community from the early 1990s. Universal interest of the concept of transparency has occurred in parallel with the globalization of the international community. Globalization can be summarized to be "rapidity and multiplicity of way in which 'here' is hostage to 'there'."[25] It promotes aspirations, values and fashions that transcend borders and cultures. As the world enters into the 21st century, this trend of globalization is not only irreversible but also reinforced, leading to the growing importance of increasing policy transparency. Transparency is both a cause and effect of lowering national borders and bureaucratic hurdles.

Definitions of Transparency

This section will present a set of definitions that have appeared in

---

[25] Rajan Menon and S. Enders Wimbush, "Asia in the 21st century: power politics alive and well," *The National Interest*, Spring 2000, p. 84.

various disciplines to illustrate how the concept of transparency has been formulated in the various fields.

1. Boutros Boutros-Ghali, 1992:[26] Openness and transparency are crucially important as part of the process of building confidence. Their significance must be emphasized, particularly at regional and sub-regional levels, in order to make military behavior more predictable and to reassure concerned states of the non-threatening intentions of potential rivals. Openness and transparency can also be useful early-warning instruments in the process of preventive diplomacy.

2. Patricia McFate, *et al.*, 1994:[27] Transparency is the voluntary or involuntary, formal or informal sharing of information that makes an event, activity or pattern of behavior more clear, open and predictable.

3. Antonia Chayes and Abram Chayes, 1994:[28] Transparency is the availability and accessibility of information about the regime and the performance of parties under it.

4. Antonia Chayes and Abram Chayes, 1995:[29] Transparency is the availability and accessibility of knowledge and information about (1) the meaning of norms, rules, and

---

[26] Boutros Boutros-Ghali, *New Dimensions of Arms Regulation and Disarmament in the Post-Cold War Era*, Report of the Secretary-General, United Nations A/C 1/47/7, October 27, 1992, pp. 18-19.

[27] Patricia McFate, *et al.*, *The Converging Roles of Arms Control Verification, Confidence-Building Measures, and Peace Operations: Opportunities for Harmonization and Synergies*, Arms Control Verification Studies No. 6 (Ottawa: Department of Foreign Affairs and International Trade, Canada, 1994), p. 81.

[28] Antonia Chayes and Abram Chayes, "Regime architecture: elements and principles," in Janne Nolan, ed., *Global Engagement: Cooperation and Security in the 21st Century* (Washington, D.C.: The Brookings Institution, 1994), pp. 66-67.

[29] Antonia Chayes and Abram Chayes, *The New Sovereignty: Compliance with International Regulatory Agreements* (Cambridge, Massachusetts: Harvard University Press, 1995), p. 135.

procedures established by the treaty and practice of the regime, and (2) the policies and activities of parties to the treaty and of any central organs of the regime as to matters relevant to treaty compliance and regime efficacy.

5. Joseph Pilat, 1996:[30] Transparency is no substitute for reductions in arms, but when properly applied it can be conducive to confidence building among states and helpful in alerting the global community to excessive accumulations of armaments. Thus, it could serve as another useful tool in facilitating non-proliferation efforts.

6. Nancy Gallagher, 1997:[31] Transparency is to know what other countries are actually doing when they claim to be in compliance.

7. Ann Florini, 1997:[32] Transparency refers to the provision of information by an actor about its own activities and capabilities to other actors. More and more, transparency is a norm—that is, a standard of behavior to which actors are held, one that has become increasingly entrenched in international security relations, politics, business practices, and policies of environmental protection.

8. Ronald Mitchell, 1998:[33] Promoting transparency is to foster the acquisition, analysis and dissemination of regular, prompt, and accurate regime-relevant information.

9. Japan Nuclear Cycle Development Institute, 1999:[34]

---

[30] Joseph Pilat, "Arms control, verification and transparency," in Jeffrey Larsen and Gregory Rattray, eds., *Arms Control Toward the 21st Century* (Boulder: Lynne Rienner, 1996), pp. 91-92.

[31] Nancy Gallagher, "The politics of verification: why 'how much?' is not enough," *Contemporary Security Policy*, August 1997, p. 139.

[32] Ann Florini, "A new role for transparency," *Contemporary Security Policy*, August 1997, p. 51.

[33] Ronald Mitchell, "Sources of transparency: information systems in international regimes," *International Studies Quarterly*, Vol. 42, 1998, p. 109.

[34] Toshiro Mochiji, *et al.*, "Joint DOE-PNC research on the use of transparency in support of

Transparency is the effort to promote mutual trust, improve credibility and establish working relationships between countries, international agencies, other nuclear entities and citizens through the sharing of information with respect to nuclear activities, both in the areas of nuclear disarmament and the peaceful use of nuclear energy.

10. Los Alamos National Laboratory, 1999:[35] Transparency is the voluntary release of information for the purpose of reassuring outside parties that one is engaging only in announced activities.

11. Neal Finkelstein, 2000:[36] Transparency is used to describe those policies that are easily understood, where information about the policy is available, where accountability is clear, and where citizens know what role they play in the implementation of the policy. Transparent policies are better than those that are opaque in terms of fairness, equity, and the democratic process.

Functions of Transparency

As a condition under which the relevant information is available to all participants, transparency allows three important functions and creates "compliance":[37] dynamic (1) to permit *coordination* between actors making independent decisions, (2) to provide *reassurance* to actors cooperating or complying with the norms of

---

nuclear non-proliferation," *Journal of Nuclear Materials Management*, Fall 1999, p. 47.

[35] Ibid., p. 48.

[36] Neal Finkelstein, "Introduction: transparency in public policy," in Neal Finkelstein, ed., *Transparency in Public Policy: Great Britain and the United States* (London: Macmillan Press, 2000), pp. 1-2.

[37] Antonia Chayes and Abram Chayes, "Regime architecture: elements and principles," p. 81.

the regime that they are not being taken advantage of, and (3) to exercise *deterrence* on actors contemplating noncompliance or defection. Analytically, the three functions may be separated but in practice, they interact and reinforce each other.

Coordination
Coordination can be tacit or explicit. In the tacit coordination, the parties have a common interest and harmony in achieving a mutually beneficial objective. There is little incentive for maximizing the difference of their payoffs—namely, the relative gain. In the explicit coordination, a normative framework consisting norms and rules is needed to regulate parties' behaviors and achieve a mutually profitable payoff. There are several examples:[38]

- International Transport and Communication: Operational and safety rules promulgated by the International Civil Aviation Organization (ICAO) are readily accepted by all parties. The Universal Postal Union has been one of the oldest international treaty regimes and operated through two major world wars for more than a century. Since 1963, the United States and the Soviet Union had established and periodically updated a direct hot line between the two Capitals in case of emergency.
- Communication Satellites: The International Telecommunication Union (ITU) system allocates for each country orbital slots for communication satellites. Before a satellite is placed in orbit, the country is required to record the slot to be occupied in a registry

---

[38] Antonia Chayes and Abram Chayes, *The New Sovereignty: Compliance with International Regulatory Agreements*, pp. 137-142.

established by the ITU. The International Telecommunication Convention also prohibits harmful interference between transmissions of member states' satellites.

- Port Control: The Memorandum of Understanding on Port State Control was signed by fourteen European states in 1982. The memorandum obliged each party not to inspect greater than 25 percent of the ships entering its harbors for compliance with various safety and environmental regulations of the International Maritime Organization (IMO). A ship that is found free of violations cannot be inspected again within six months. The parties also notify each other, through a computerized network, of the inspected ships and inspection results.

- Arms Trade Disclosure: The UN Register of Conventional Arms is one of the most ambitious efforts to achieve coordination by information dissemination. The Register was established by a General Assembly resolution calling on member states to voluntarily record their sales of conventional armaments of certain categories—battle tanks, armored combat vehicles, large caliber artillery systems, combat aircrafts, attack helicopters, warships, missiles and missile launchers.

Reassurance

Reassurance can be produced by the creation of a normative framework itself with little compliance considerations. Or it can be generated by monitoring of compliance behavior of member states of a treaty regime. In the former, verification is of little concern because incentives to violate an agreement are low. The examples are as follows:[39]

- Various arms control treaties with virtually no military significance can create reassurance just by their existence. The 1957 Antarctic Treaty, 1967 Outer Space Treaty, and the 1971 Seabed Treaty aim to prohibit using respective regions for military purposes and deploying weapons of mass destruction there. These treaties, although with little military significance, have served as legal constraints to block individual attempts to militarize the respective areas.

- Some international treaty regimes focus on providing various assistances for inducing cooperation rather than on detecting violations. The Montreal Protocol for the Protection of the Ozone Layer provides technical assistance to help less developed parties to adhere to the treaty regimes. It was amended in 1990 to give financial assistance to developing countries for their compliance with the regimes. In specific, the article 10 of the London Amendments undertook to cover the agreed incremental costs of compliance for developing countries. And an entire chapter of Agenda 21, adopted at the 1992 Earth Summit at Rio, deals with national mechanisms and international cooperation for building technical capacity in developing countries. The Framework Convention on Climate Change (FCCC) created a fund of 2 billion U.S. dollars that is managed by the Global Environmental Facility (GEF). The GEF is supposed to use this money to enable activities undertaken by developing countries such as planning, institutional strengthening, training,

---

[39] Ibid., pp. 142-151.

research and education that will facilitate effective implementation of the Convention.

In the latter case of reassurance, the importance of verification is highly appreciated because incentives for treaty violations are high. When the incentives to violate an agreement are strong or the costs of defection by the other party are high, more is needed for reassurance than mere existence of treaty regimes. Detail information about the compliance behaviors of the other parties is demanded, in which transparency becomes a key to provide reassurance.

- As arms control treaties become militarily significant, so they come to have very complex sets of verification mechanisms. In case of the 1987 Intermediate-range Nuclear Forces (INF) Treaty, verification provisions cover one-third of the whole treaty. The 1991 Strategic Arms Reduction Talks (START) Treaty may have equipped the most voluminous verification provisions in the history of verification. Out of 280 pages of the treaty, the parts covering verification and related provisions amount to 181 pages—65 percent of the treaty. The 1993 Chemical Weapons Convention (CWC) also has an extensive verification system reaching out to civilian sectors as well as military programs due to dual-use nature of most chemical agents.

Another factor that influences the degree of verification is the level of trust between treaties parties. The higher the level of trust is, the lower the degree of verification would be. And the lower the level of confidence is, the tighter the verification system would be. The best example is the American demand against

North Korea of "CVID" as a principle to resolve the North Korea nuclear crisis. The CVID means complete, verifiable, and irreversible dismantlement of North Korea's nuclear program. This principle was derived from the United States' strong distrust of North Korea as a reliable negotiating partner. The U.S. mistrust against North Korea was reinforced by Pyongyang's persistent violations of the Geneva Agreed Framework signed in October 1994.

- "Complete" is meant not to repeat the mistake of the Agreed Framework that stopped short of achieving complete transparency of the North's nuclear history and complete dismantlement of its nuclear capability and infrastructure. John Bolton confirmed that the highly enriched uranium (HEU) program as well as the plutonium program should be eliminated to attain the goal of complete dismantlement.[40] Regarding the HEU program, North Korea has said that it is willing to discuss technical matters if the United States presents related evidence.[41] On the other hand, it is the U.S. position that providing evidence about the HEU program—which is much easier to hide than the plutonium program—would only help North Korea's concealment activities.[42]

---

[40] *The Bush Administration's Non-proliferation Policy: Successes and Future Challenges,* Testimony by Under Secretary of State for Arms Control and International Security John Bolton to the House International Relations Committee, March 30, 2004.

[41] During a U.S. delegation's visit to North Korea in January 2004, Vice Minister Kim Gye Gwan said that North Korea had no facilities, equipment or scientists dedicated to an HEU program and added that "we can be very serious when we talk about this. We are fully open to technical talks." Siegfried Hecker, Senate Committee on Foreign Relations Hearings on Visit to the Yongbyon Nuclear Scientific Research Center in North Korea, January 21, 2004, p. 10.

[42] Under Secretary of State for East Asia and Pacific Affairs James Kelly's response to a question by Senator Richard Lugar at the Senate Foreign Relations Committee Hearing,

- "Verifiable" is based on the Bush administration's deep-seated mistrust of North Korea and manifests its will to set up a reliable verification mechanism to effectively monitor the North's compliance behavior.
- "Irreversible" is meant to eradicate all human, materials and technical infrastructures for nuclear development programs and in consequence, making it impossible for such programs to be revived in the North Korean territories. According to John Bolton, irreversible dismantlement attains its goal when North Korea abandons both its so-called "civil" and "peaceful" nuclear programs as well as military programs and permits the removal of all critical components.[43]

In short, the agreement plus the continuing assurance that the other parties are complying will do the trick, and resultant transparency will provide reassurance.

Deterrence

By increasing transparency of the parties' compliance behaviors, the treaty participants are effectively deterred from breaking away from the treaty. Deterrence was the most important strategic paradigm governing the international security relations during the Cold War, and the most visible manifestation of which was the Mutual Assured Destruction—MAD. MAD was a form of deterrence developed at the early stage of the Cold War. The basic tenet of MAD was to deter the other side's attack by maintaining capability and threat to retaliate the opposite side to a degree to wipe out the effect of the attack. That is, country A would not

---

March 2, 2004, http://www.ifins.org/pages/ kison-archive-kn545.thm.
[43] *The Bush Administration's Non-proliferation Policy: Successes and Future Challenges.*

attack country B if A knew that B could in retaliation inflict unacceptable damage on A, regardless of the nature or timing or duration of the initial attack.[44]

From the perspective of transparency and information exchange, deterrence is the obverse of reassurance and each acts at the opposite end of the transaction. This mutual relationship between deterrence and reassurance is detailed as follows:[45]

> A party disposed to comply needs reassurance. A party contemplating violation needs to be deterred. Transparency supplies both. The probability that conduct departing from treaty requirements will be discovered operates to reassure the first and to deter the second, and that probability increases with the transparency of the treaty regime. The efforts of treaty organizations to provide information about the compliance of members thus have a deterrence as well as a reassurance effect.

In conclusion, coordination mostly occurs when norms are institutionalized while reassurance works in most cases only if a system of verifying the compliance behavior is set in motion. Deterrence and reassurance are similar to two sides of a coin and promote compliance of an agreement. In practice, coordination, reassurance and deterrence are interwoven to enhance transparency.

Global Phenomena of Increasing Policy Transparency

Transparency is spreading as part and parcel of three trends:

---

[44] Coit Blacker and Gloria Duffy, *International Arms Control: Issues and Agreement* (Stanford, CA: Stanford University Press, 1984), p. 203.

[45] Antonia Chayes and Abram Chayes, *The New Sovereignty: Compliance with International Regulatory Agreements*, p. 151.

democratization, globalization and dramatic advances in technology. Increasingly, in issues ranging from security to commerce to economics, transparency is the preferred means of enforcement. In fact, the international community is embracing new standards of conduct enforced by willful disclosure: "regulation by revelation."[46]

In the realm of security, the best example is that the verification provisions in various arms control treaties have been strengthened. First of all, the IAEA adopted the additional protocol to reinforce the existing safeguards provisions and to close loopholes for possible violations. According to the additional protocol, the following measures are taken among others: (1) more detail information are disclosed including research plans, nuclear facility design, and dismantled facilities; (2) the IAEA can inspect not only nuclear-related facilities but also research activities and facilities that can be used for nuclear development; (3) environmental sampling is allowed; and (4) remote monitoring devices are permitted. And it is outstanding that the arms control treaties agreed in parallel with the dissolution of the Cold War had encompassed greater transparency. For instance, the 1990 Conventional Forces in Europe I (CFE I) Treaty, the 1992 Strategic Arms Reduction Talks (START) Treaty, the 1992 UN Register of Conventional Armaments, and the 1992 Open Skies Treaty involve revealing extensive amounts of information voluntarily and involuntarily.

In particular, it is noted that the 1993 Chemical Weapons Convention (CWC) has established an elaborate, powerful, and intrusive structure to monitor all production and acquisition of a

---

[46] Ann Florini, "The end of secrecy," *Foreign Policy*, Summer 1998, pp. 53-60.

variety of chemicals by its member states. This leads to a significant reduction in the information asymmetry and resultant increase in transparency of member nations' policies related with chemical weapons. The CWC prohibits development, production, acquirement, store, transfer, and use of chemical weapons and demands destruction of existing weapons and production facilities. Entered into effect in 1997, the CWC is the second international arms control treaty to inspect the civilian industries as well as military installations after the NPT. The CWC is equipped with an elaborate and comprehensive system of verification, which inflicts civilian industries as well as military establishments of member states.

The Organization for the Prohibition of Chemical Weapons (OPCW) is an administrative and inspection body to implement the CWC. Each member state is obliged to report its chemical weapons capability and the weapons should be dismantled under the supervision of the OPCW. Each party also should report production activities and inventories of dual-usable chemical material regardless of whether they are owned by private companies or the government. The OPCW can conduct regular inspections at private or governmental chemical facilities and has the right of a special inspection with 12 hours of pre-notification period.

There are protective measures against member states' compromising their commercial interests by adhering to the CWC. For example, inspection procedures for commercially sensitive places are specified according to the *managed access* principle so as to protect commercially valuable information. To prevent special inspections from being misused in commercial competition, the CWC executive council can annul a special inspection request of member states with the approval of 75% of the whole member states. The CWC demonstrates that, at least in principle, it is possible to design a regime that contains credible end-use control procedures and is also able to deal with the sensitive issue of proprietary information.[47]

Although the CWC and the NPT both cover civilian industries as well as military installations, the CWC would have much greater impacts on the civilian sectors mainly because of the sheer size of the civilian chemical industries around the world. Although the NPT and the IAEA have 189 and 138 member states, respectively, and more than 140 countries separately signed a safeguard agreement with the IAEA as of 2005, there are only a couple of dozens countries where nuclear power is a main energy source of that country.[48] In case of the CWC, it is obvious that a number of countries in the world have chemical industries of sizable scales. In this respect, the verification provisions of the CWC would have across-the-board and worldwide influences on governments' policies and civilian industrial sectors.

Like the NPT, the CWC begins a verification procedure by assuring whether the initial report by a member state on its chemical weapons and/or industrial programs confirms to the reality in the field. In this process, such sensitive issues are raised due to dual-use nature of most chemical agents as protecting confidentiality of intellectual properties and maintaining equity among the member countries in terms of obligations and rights.

---

[47] Wolfgang Reinicke, "Cooperative security and the political economy of non-proliferation," in Janne Nolan, ed., *Global Engagement: Cooperation and Security in the 21st Century* (Washington, D.C.: The Brookings Institution, 1994), p. 213.

[48] As of 2005, a total of 30 countries operate at least one nuclear power plant but only nine countries have more than 9 such plants.

The CWC has been successful to manage these issues. In relation to transparency, efforts to protect intellectual properties had been made from the beginning of treaty negotiations.[49]

First, considering the extensive dual-uses of chemical agents, the treaty negotiating countries listened to various opinions of private sectors by using so-called "horizontal subsidiarity." In seeking to improve the legitimacy, acceptability, efficiency, and effectiveness of public policies, horizontal subsidiarity delegates some of public policymaking to non-state actors—for example, to businesses and their associations, labor groups, non-governmental and non-profit organizations, consumer groups, and other interested parties in the country.[50] Each of the negotiating countries had separate discussions with its private chemical industries, and in 1989 a Government-Industry Conference against Chemical Weapons was held, which brought together public officials and civilian industries from 60 countries. The chemical companies of these countries represent 95 percent of the world's chemical production capacity.

Second, as in the IAEA, the CWC also prohibits any information obtained through verification from being released in public. For example, the CWC has the confidentiality annex in the Convention text, which provides detail obligations and guidelines for staffs and processes. The OPCW that governs the CWC Executive Council and the Technical Secretariat is subject to detail requirements relating to access, treatment, and storage of information at various levels of confidentiality. The CWC also

---

[49] Wolfgang Reinicke, *Global Public Policy: Governing without Government?* (Washington, D.C.: The Brookings Institution, 1998), pp. 210-214.
[50] Ibid., p. 89.

provides detail guidelines for the employment and conduct of personnel in the Technical Secretariat. For example, each staff member should maintain secrecy on what he learned while working for the OPCW for five years after termination of employment. Employee performance evaluations must give specific attention to the employees' record in protecting confidential information. And process-oriented provisions of the CWC allow facilities to be inspected to take necessary confidentiality precautions and establish stringent guidelines for the transport and analysis of samples taken. Similarly, inspectors are required to conduct inspections in the least intrusive manner.

Chapter Three

Non-Proliferation and Transparency

in the 21st Century

Following the end of the Cold War, the international security environments have transformed to a great extent that has made it necessary to change basic conditions of formulating national security strategies. Disintegration of the Soviet Union, collapse of the Eastern communist block, drastic reduction of American and Russian nuclear weapons,[51] conventional force reductions in Europe, and the Chemical Weapons Convention, etc. are the positive developments for stability and peace of the world. On the other hand, deep-rooted historic enmity, racial conflicts, political and/or territorial disputes, regional arms competition, proliferation of weapons of mass destruction—WMD—and missiles, and emergence of asymmetric security threats are negative developments damaging regional security and international peace. Particular

---

[51] On May 24, 2002, the United States President George Bush and Russia President Vladimir Putin signed the Strategic Offensive Reductions Treaty (SORT) and agreed to reduce their strategic nuclear warheads up to 1,700-2,200 by the end of 2012. The text of the treaty can be found at http://www.clw.org/control/sort/treatytext.html. The SORT is also called Moscow Treaty on Strategic Reductions.

attention is paid to growing threats caused by proliferation of WMD and its related technologies among the developing countries.

At the same time, awareness of the significance of enhancing policy transparency has been spread throughout the world. To increase transparency in the course of formulating and implementing policies has been settled as a new norm in the various areas of politics, economy, security, finance, and environment, etc. The following observation succinctly describes values of upholding policy transparency:[52]

> Transparency has a practical, perhaps even a militant, utility. These days, openness is the one theme we can assertively promote worldwide without apology or liability. On the one hand, it seizes high moral ground, from almost any cultural perspective. On the other, it stymies potential foes while helping make the world a safer place.

Transparency is not a new concept to the South Korean public. Among the many reasons that caused the economic crisis in the late 1990s, the international financial institutions pointed out an inappropriate level of transparency in the governmental and civilian financial policies of the ROK. As South Korean society becomes democratized, transparency has been increasingly embodied as an important norm in South Korean domestic politics. Nowadays, major offices in central and local governments must reveal the amount of public service account and the list of its uses. Non-governmental organizations often disclose unknown information about candidates of parliamentary elections, local elections and ministerial postings, thus exercising significant influences on election results and nominations.

---

[52] David Brin, "Letters: transparency's virtues," *Foreign Policy*, Fall 1998, p. 173.

Cooperative Security and Transparency

Since the end of the Cold War, the two distinct characteristics are noted in the realm of international security: (1) the expansion of threat origins and (2) the diversity and lethality of threat nature. The first reflects the two simultaneous phenomena. On the one hand, the international security was freed from the bipolar structure that had dominated global geopolitics based on the United States and the Soviet Union during the Cold War. On the other hand, it is observed that historical, racial, and cultural animosities have been resurfacing above the water, triggering security uncertainties and military tensions in the various parts of the world. In addition, non-state actors such as terrorist organizations as well as rogue states are becoming growing threats to international peace.

The second mainly reflects a worldwide phenomenon that WMD and related technologies are rapidly spreading around the world. Globalization has made it more convenient to have frequent exchanges of ideas, technologies, information, and people themselves. The revolutionary advancement of information technologies must have contributed greatly to this unprecedented flow of exchanges. WMD proliferation is not an exception to this new reality. Diversity of weapons—chemical, biological, nuclear and radiological ones—and their critical lethality bring about serious concerns of the international community especially after the 9/11 terror incident. In response to emerging threats from the 1990s, the international society has gathered wisdom and efforts of all concerned nations to further strengthen international non-proliferation norms, which will be discussed at the following section.

The basic philosophy of cooperative security is that the appropriate principle for dealing with these new security threats is that of cooperative engagement—"a commitment to regulate the size, technical composition, investment patterns, and operational practices of all military forces by mutual consent for mutual benefits."[53] The central aim of cooperative security is to seek to devise agreed-upon measures to avoid war primarily by preventing the means for successful aggression from being assembled, thus obviating the need for the threatened nations to make their own countermeasures.[54] Cooperative security recognizes and tries to articulate how the character of international security has changed since the end of the Cold War and to demonstrate how this change has rendered the foundations of strategy used during the Cold War no longer appropriate.[55] In practice, cooperative security attempts to replace preparations to counter threats with the prevention of such threats in the first place, and replaces the deterring of aggression with actions to make preparation for it more difficult.[56]

The differences between collective security during the Cold War and cooperative security in the post-Cold War era are distinctive. In collective security, a nation deters the other from using military forces and defeats it if deterrence fails; contains and confronts expansion of communism; forms military alliances; and maintains secrecy and ambiguity of major policies. On the other hand, in cooperative security, a nation uses multilateral sanctions

---

[53] Ashton Carter, William Perry and John Steinbruner, *A New Concept of Cooperative Security* (Washington, D.C.: The Brookings Institution, 1992), p. 6.

[54] Ibid., p. 7.

[55] Janne Nolan, "The concept of cooperative security," in Janne Nolan, ed., *Global Engagement: Cooperation and Security in the 21st Century* (Washington, D.C.: The Brookings Institution, 1994), p. 5.

[56] Ibid.

and inducements to prevent the other from equipping with a large-scale attack capability; pursues a cooperative engagement; reinforces multilateral security cooperation and confidence building; and enhances disclosure and transparency of major policies. Assuming fundamental security objectives can be compatible in the multilateral context, cooperative engagement seeks to attain its security objective through institutionalized consent rather than through threats of material or physical coercion and pursues to establish collaborative rather than confrontational relationships among nations.[57] Cooperative security differs from collective security much as preventive medicine differs from acute cure.[58]

It has been said that there exist five ingredients of cooperative security.[59] The first is to restrain uses and utilities of nuclear weapons. In this respect, cooperative security puts emphasis on forbidding emergence of a new nuclear weapon state, blocking modernization of nuclear weapons, and carrying out nuclear disarmament. At the same time, the function and role of nuclear weapons are minimized. Of course, it is noted that some of Bush administration's nuclear-related policies seem contrary to reducing the role of nuclear weapons. For example, it has moved to develop very small-yield nuclear weapons for the mission of earth penetration and bunker busting.[60] It wants to maintain the

---

[57] Ibid., p. 4.

[58] Ibid., p. 5.

[59] Ashton Carter, William Perry and John Steinbruner, *A New Concept of Cooperative Security*, pp. 11-41.

[60] The Nuclear Posture Review (NPR) submitted to the Congress on December 31, 2001 proposed to develop earth-penetrating weapons (EPW) as a solution to increase striking capability for the hard and deeply buried target (HDBT). The text of the NPR can be found at http://www.globalsecurity.org/wmd/library/policy/dod/npr.htm. Yield of the EPW is likely to be around 5,000 pounds. William Arkin, "Secret plan outlines the unthinkable," *Los Angeles Times*, March 10, 2002.

right of testing nuclear weapons, thus refusing to ratify the Comprehensive Test Ban Treaty. These policies, however, have faced strong objections within and without the United States and created political burdens for the Bush administration. At the 7th NPT Review Conference in May 2005, American lukewarm attitudes on nuclear arms control were under severe criticism by many participating states.

The second ingredient of cooperative security is to make postures of conventional military forces defensive by tightening the restraints of offensive military capabilities such as tactical airplanes and offensive ground forces. This was the underlying philosophy of the Conventional Forces in Europe (CFE) Treaties. The third one is to react to any aggression multilaterally by calling on the international organizations such as United Nations for appropriate actions. For instance, the Gulf War and conflicts in Bosnia and Kosovo were all managed by multilateral cooperation. The fourth ingredient is to deter WMD from being proliferated worldwide. International awareness of the dangers of WMD proliferation and subsequent cooperation to crack down proliferation attempts from non-state as well as state actors have been getting stronger ever. Finally, cooperative security requires a nation to enhance transparency of its military policy in order for the others to confirm its military postures and intentions. In this context, cooperative security puts emphasis on the significance of intrusive monitoring and tries to reassure that no agreement is being violated. In addition, incentives are provided to induce compliance and sanctions are imposed if a violation occurs.

Transparency is a principal element for promoting cooperative security. Specifically, it is a practical tool to embody the strategic principle of cooperative security—cooperative engagement—into reality. Florini proposes three motivations that make transparency flourish: technology development, democratization, and globalization.[61] Dramatic advances in technology have made transparency more feasible and attractive. With the spread of democracy, powerful entities such as states and big corporations are likelier to be held accountable for their behaviors. And as the world shrinks, a lot of people come to have a desire to know and have a say in what used to be not of their concern.

Nolan presented five underlying factors that make cooperative security imperative.[62] These factors can be regarded as what makes transparency necessary and important in the first place. The first factor is the diffusion of civil and military technology. As scientific developments proceed, more countries come to possess sensitive and dual-use technologies. Under this circumstance, technology cooperation among nations promotes proliferation of such technologies worldwide, making it necessary to have such cooperation more transparent to prevent circumvention of international non-proliferation norms. The second factor is shrinking military budgets and export markets caused by the drastic reduction of military capabilities in Europe in the 1990s. Many nations have tried to cushion the economic and social impact of the shrinkage by increasing their sales of weapons and technologies to other nations, often those in the politically troubled parts of the world. Competition of exporting countries, provision of sensitive technologies in arms trade arrangements, and increase of civil firms to exploit dual-use technologies are contributing to proliferation of dangerous

---

[61] Ann Florini, "The end of secrecy," pp. 52-53.

[62] Janne Nolan, *et al.*, "The imperatives for cooperation," in Janne Nolan, ed., *Global Engagement: Cooperation and Security in the 21ˢᵗ Century* (Washington, D.C.: The Brookings Institution, 1994), pp. 20-45.

weapons and technologies, which highlights the importance of increasing transparency of exporting policies and sharing of information among arms supplying nations.

The third factor is the internationalization of economic activity. Breakdown of trade barriers, active foreign investment, real-time international trade, and vigorous exchanges of information are making economic activity more internationalized. Riding on this trend, the diffusion of dual-use technologies is also accelerated. According to Wolfgang Reinicke, the internationalization of economic activity and the resultant technology diffusion have led to globalization of the defense industry and "internationalization of availability" of technologies.[63] As the available technologies expand in a free and diverse manner, the importance grows of keeping national policies of trading dual-use goods and technologies transparent.

The fourth factor is the disintegration of the Soviet Union. A negative but serious repercussion from the demise of the Soviet Union is the intentional and uncontrolled proliferation of WMD, related technologies and knowledge to the dangerous regimes and terrorists—an issue of smuggling and brain drain. Most recently, the revelation about the nuclear smuggling network of Abdul Qadeer Khan, the father of Pakistan's nuclear weapons, has alarmed the international community about the necessity to tighten a nation's domestic controls of scientists and technologies, to make its policies more transparent, and to strengthen the existing non-proliferation regimes.

---

[63] Wolfgang Reinicke, *Global Public Policy: Governing without Government?*, p. 177.

The final driving force for cooperative security and policy transparency is the resurfacing of regional disputes shadowed by the superpower confrontation during the Cold War. Freed from the U.S.-Soviet bipolar structure, historical, territorial, racial, and religious disputes have spread around the world. In order to prevent regional arms race, minimize unnecessary tension, and remove possible misunderstandings, international norms and regimes should be reinforced to curb proliferation of WMD and related policies of concerned nations must be transparent as much as possible.

Efforts to Strengthen Non-Proliferation Regimes

The 7th NPT Review Conference was held in New York from May 2 to 27, 2005. Among the 189 member states, 150 countries dispatched their delegations. In the midst of dangerous events since the beginning of the 21st century—such as the 9/11 tragic incident, ever intensified North Korea nuclear crisis, and Iran's secret nuclear program, the Review Conference has drawn keen attention worldwide as well as from the non-proliferation community. In light of intentional attention paid to the Review Conference, however, the result was disappointing. The final document was not adopted, signifying very little was achieved in the Conference. But it is premature to bemoan the collapse of the global non-proliferation regime itself. If the failure of the Review Conference reinvigorates concern about the future of the NPT, according to one positive view, then the failure to produce a final document might be a price worth paying.[64] In fact, there are positive developments in the Review Conference. For instance,

---

[64] Brad Glosserman, "The current situation in the field of nuclear disarmament and non-proliferation," a paper presented at the International Symposium on *Peace and Environmental Issues*, held in Kanazawa city, Japan, June 13-14, 2005.

states parties confirmed that the NPT has great significance for maintaining international peace and security and that compliance with the treaty is extremely important, emphasizing the values of IAEA safeguards agreement and additional protocol. In addition, a variety of proposals were submitted during the Conference to contribute to strengthening the NPT regime and intensive exchanges of views were made regarding the proposals. They should provide valuable sources for future endeavor on strengthening the international non-proliferation regime and nuclear disarmament.[65]

Despite regrettable outcome of the 7th NPT Review Conference, it is noted that an international trend has awakened many nations of the danger of WMD and driven them to strengthen the non-proliferation regimes to prevent further proliferation. The Gulf War of 1991 provided a striking demonstration of the growing risks associated with WMD proliferation. Iraq's covert nuclear weapon development program in violation of the NPT and the IAEA safeguards agreement was revealed by the United Nations Special Commission (UNSCOM), which was a major force that drove the United Nations to take an alarming position on the danger of WMD proliferation. Since January 1992, in particular, the United Nations has defined that proliferation of nuclear, chemical and biological weapons as well as their means of delivery constitutes a threat to international peace and security. At the summit meeting of the Security Council at the Level of Heads of States and Government held on January 31, 1992, the members of the Security Council pledged "to commit themselves to

---

[65] Takeshi Nakane, "Japans' efforts in disarmament and non-proliferation after the 2005 NPT Review Conference," a paper presented at the International Symposium on *Peace and Environmental Issues*, held in Kanazawa city, Japan, June 13-14, 2005.

working to prevent the spread of technology related to the research for or the production of such weapons and to take appropriate action to that end."[66] Thereafter, a number of important steps have been taken, on international, regional, national bases, in order to curb proliferation of WMD, missiles, related materials and technologies, and terror activities including the following important measures.

The IAEA Actions

Realizing the potential danger of nuclear fissionable and radioactive materials, the IAEA has long made great contributions to securing these materials dispersed around the world.[67] For example, the general conference of the IAEA in September 1994 adopted a resolution calling on its members to take all necessary measures to prevent illicit trafficking of nuclear material. In December the same year, the director general of the agency called for other radioactive sources to be dealt with in a similar fashion. In 2001, the agency established a nuclear security fund amounting to 23 million U.S. dollars. The purpose is to assist member states in locating and securing radioactive sources, detecting nuclear smuggling, and establishing national regulatory oversight bodies and national source registries. The IAEA has also maintained an Illicit Trafficking Database and has fostered a Code of Conduct on the Safety and Security of Radioactive Sources. Contrary to the international trends, North Korea is not a member of the IAEA, which makes it impossible for the Agency

---

[66] Refer to the section on "Disarmament, arms control and weapons of mass destruction," in the *Statement of the Security Council at the Level of Heads of States and Government*, United Nations Security Council, S/23500, January 31, 1992.

[67] Klaas van der Meer, "The radiological threat: verification at the source," *Verification Yearbook 2003* (London: The VERTIC, 2003), p. 126.

to demand the country to accept its guidelines.[68]

The Cooperative Threat Reduction Program
The United States initiated the Cooperative Threat Reduction (CTR) Programs to dismantle WMD in the former Soviet Union and convert other dual-usable military capabilities for peaceful uses where possible. After the disintegration of the Soviet Union, it had emerged as a top U.S. security concern to safeguard the Soviet nuclear weapons deployed at the four Republics—Belarus, Kazakhstan, Russia and Ukraine—and to prevent nuclear weapons, materials, equipments, and scientists from flowing out of the Republics. Based on the initiative of Senators Sam Nunn and Richard Lugar, the U.S. Congress established the Nunn-Lugar Cooperative Threat Reduction Programs in November 1991. The programs focus on four key objectives:[69]

- Destroy nuclear, chemical, and other weapons of mass destruction;
- Transport, store, disable and safeguard these weapons in connection with their destruction;
- Establish verifiable safeguards against the proliferation of these weapons, their components, and weapons-usable materials; and
- Prevent the diversion of scientific expertise that could contribute to weapons programs in other nations.

---

[68] North Korea withdrew from the IAEA membership on June 14, 1994 when a U.N. sanction was pursued against its taking out spent fuels from the 5MWe reactor in May 1994.

[69] *Cooperative Threat Reduction* (Washington, D.C.: U.S. Department of Defense, April 1995), p. 4.

For example, in order to keep nuclear experts, technologies and materials from flowing out abroad, scientific research centers were established to hire nuclear scientists and technicians.[70] In 1992, the International Science and Technology Center (ISTC) opened in Moscow and as of November 2000, about 30,000 scientists from 400 research institutes in the four Republics were working on 1,156 projects at the cost of 316 million U.S. dollars. In 1995, the Science and Technology Center of Ukraine (STCU) was established and as of mid-2000, around 6,700 scientists were participating in 290 projects and 42 million U.S. dollars were expended.

In the field of dismantling nuclear capability, the CTR programs have made significant achievements.[71] By the end of 2000, U.S. Department of Defense had deactivated 5,288 missile warheads, destroyed 419 long-range nuclear missiles and 367 silos, eliminated 81 bombers, 292 submarine missile launchers and 174 submarine missiles, and sealed 194 nuclear test holes and sites. U.S. Department of Energy decided to buy 500 metric tons of highly enriched uranium, an equivalent of 25,000 warheads, and convert them to low enriched uranium that can be used as commercial fuel in nuclear reactors. As of 2001, 100 metric tons of HEU was purchased and converted. As of 2001, the CTR programs had spent about six billion dollars.[72]

---

[70] *Nuclear Status Report: Nuclear Weapons, Fissile Material, and Export Controls in the Former Soviet Union* (Monterey: Monterey Institute of International Studies, June 2001), pp. 68-74.

[71] Vladislav Nikiforov, "U.S. reviewing aid for non-proliferation programs in Russia," April 17, 2001, http://www.bellona.no/imaker?id=20093&sub=1.

[72] "Effective nuclear disarmament," *New York Times*, March 31, 2001.

The Proliferation Security Initiative

The Proliferation Security Initiative (PSI) was initiated by the Bush administration in May 2003. Not being just a political rhetoric or a diplomatic campaign, the PSI is a coercive strategy of concerned countries to prevent rogue regimes and terrorist groups from acquiring WMD and financial sources for such purposes. The PSI officially started on June 12, 2003 when eleven core members gathered in Madrid, Spain to discuss ways to implement the PSI initiative nationally and internationally.[73] As part of the PSI, in September 2003, Australia, Britain, Japan and the United States carried out the first joint military exercise at the Coral Sea off Australia to train for the interception of ships to and from nations suspected of having illegal weapons programs.[74]

According to John Bolton, the objective of the PSI is not just to prevent the spread of WMD but also to eliminate or "roll back" such weapons from rogue states and terrorist groups.[75] While pursuing diplomatic dialogues, the PSI is willing to deploy more robust tactics such as economic sanctions, interdiction and seizure, and even preemptive strike where required. The PSI calls for international community to take aggressive measures to root out existent and potential capabilities to develop and spread WMD from countries like North Korea. Thus, the basic concept of the PSI conforms to the principle of CVID regarding North

---

[73] The eleven core members are Australia, Britain, France, Germany, Italy, Japan, the Netherlands, Portugal, Poland, Spain and the United States. In June 2004, Russia joined PSI as a new core member.

[74] Steven Weisman, "U.S. plans more sea exercises on halting illegal arms trade," *New York Times*, September 10, 2003.

[75] *Testimony of John Bolton, Under Secretary for Arms Control and International Security, U.S. Department of State*, Committee on International Relations, United States House of Representatives, June 4, 2003, http://www.house.gov/international_relations/108/bolt060 4.html.

Korea's nuclear capability.

Although the original motivation was to ban the proliferation of WMD, the PSI regards rogue states' illegal activities such as drug trafficking, money counterfeiting and laundering as major financial sources to support their WMD programs. John Bolton argued that "as we close off proliferation networks, we inevitably will intercept criminal activity and overlapping smuggling rings."[76] Although the PSI does not designate a specific target country, there are many incidents indicating that North Korea will be an important target of the PSI. For example, in April 2003, Australian special forces seized a North Korean freighter called *Pong Su* that allegedly delivered 50 million U.S. dollars worth of heroin. The ship was registered in the North Korean port of Nampo. In the same month, French and German authorities intercepted shipments of aluminum tubes and sodium cyanide likely bound for North Korea's nuclear and chemical weapons programs.

The UN Security Council Resolution 1540
The United Nations Security Council was determined to forcefully react to the risk and danger of proliferating WMD, related items, dual-use technologies and materials. The UN Security Council adopted the Resolution 1540 in April 2004 and expressed grave concerns with the threat of illicit trafficking in nuclear, chemical, or biological weapons, delivery means and related materials. The Security Council put particular emphasis on the danger of proliferating materials, which it argues, "add a new dimension to the issue of proliferation of such weapons and

---

[76] Ibid.

also pose a threat to international peace and security." [77] According to the Resolution 1540, all states should develop and maintain measures to account for, secure, physically protect WMD, missiles and related materials; develop and maintain border controls and law enforcement efforts to detect and combat the illicit trafficking and brokering in the weapons and materials; and develop and maintain national export and transshipment controls over such weapons and materials. According to a view, the PSI has been strengthened by the Resolution 1540. [78]

The Container Security Initiative
The United States has dispatched customs officials to major ports in the world and is conducting inspections beforehand of container ships heading toward America to check whether WMD and related materials are loaded. [79] Dubbed as the Container Security Initiative—CSI, this policy aims at ensuring safety of and preventing potential security risks from all containers to be brought into the United States. In this context, Washington began to pay attention to not only origins but also possible routes and engaged partners of proliferation. For example, John Bolton, Under Secretary of State for Arms Control and International Security, insisted that "the frontlines in our non-proliferation strategy must extent beyond the well-known rogue states to the trade routes and entities engaged in supplying proliferant countries." [80]

---

[77] *U.N. Security Council Resolution 1540*, April 28, 2004. In case of missiles, the MTCR member states already agreed, in the 1995 plenary meeting held in Bonn, to cooperate on inspection and interdiction of ships causing proliferation concerns.

[78] Dan Blumenthal, "Facing a nuclear North Korea," *Asian Outlook*, June-July 2005, p, 7, http://www.aei.org/asia.

[79] For example, U.S. customs officials are stationed in Busan in South Korea and Shanghai and Shenzhen in China.

[80] John Bolton, "An all-out war on proliferation," *Financial Times*, September 7, 2004.

The Transshipment Country Export Control Initiative
There exist growing concerns about the risks of transshipment—unloading goods from a ship and reloading them onto a different ship in a port. One response is that under the name of Transshipment Country Export Control Initiative—TECI, efforts and discussions are undergoing to institutionalize an international system to prevent and curtail transshipment-related activities. For instance, officials of 22 countries and economic regions met at the Global Transshipment Control Enforcement Conference held in Sydney from July 15-18, 2003. They discussed appropriate enforcement policies and practices to make transit, transshipment, and re-export trade less vulnerable to terrorism and exploitation of legitimate commerce by those wishing to acquire illicitly WMD, delivery means, and their related goods and technologies.[81] The conference articulated six principles of the TECI, which emphasize the importance of effective controls on the transit, transshipment, and re-export of WMD and related items and of cooperation and information sharing among governing authorities for that purpose.[82]

The European Union-American Cooperation
The European Union and the United States agreed on a joint program on pursuing the non-proliferation of WMD.[83] The two sides agreed that proliferation of WMD and their delivery

---

[81] The participant countries are Australia, Cyprus, Estonia, Fiji, Hong Kong, Indonesia, Japan, Jordan, Latvia, Lithuania, Malaysia, Malta, New Zealand, Oman, Pakistan, Panama, Singapore, Taiwan, Thailand, Turkey, the United Arab Emirates, and the United States of America.

[82] Available at http://www.bxa.doc.gov/ComplianceAndEnforcement/TECISydney7_03 Principles.htm.

[83] *Joint Statement by the European Union and United States on the Joint Program of Work on the Non-proliferation of Weapons of Mass Destruction*, June 20, 2005, http://www.whitehouse.gov.

systems continue to be a preeminent threat to international peace and security. They reaffirmed that this global challenge needs to be tackled individually and collectively and requires an effective global response. They also vowed that the United States and the European Union were steadfast partners in the fight against the proliferation of WMD and declared to undertake the following new initiatives to strengthen cooperation and coordination: (1) building global support for non-proliferation, (2) reinforcing the NPT, (3) recognizing the importance of the biological threat, (4) promoting full implementation of the UN Security Council Resolution 1540, (5) establishing a dialogue on compliance and verification, (6) strengthening the IAEA, (7) advancing the PSI, and (8) upholding the global partnership to expand the cooperative threat reduction program worldwide.

The International Convention for the Suppression of Acts of Nuclear Terrorism

On April 13, 2005, the United Nations General Assembly adopted the International Convention for the Suppression of Acts of Nuclear Terrorism—CSANT. The Convention requires member states to establish offenses in their domestic law for activities related to nuclear terrorism. It also provides a framework for international cooperation on the investigation and prosecution of nuclear terrors and for extradition of criminals. The Convention was based on the first draft submitted by the Russian Federation in 1997. The Convention calls for member states to provide each other with legal assistance to facilitate appropriate national implementation. Nuclear or radiological materials seized are subject to verification under the IAEA's health, safety and physical protection standards.

Discrimination and Double Standard

While a series of important efforts have been undertaken in order to curb proliferation of WMD as explained above, it is noted that several events have occurred that could hamper such non-proliferation efforts. These events could kindle long-held complaints about "discrimination"—the nuclear weapon states, particularly the United States, which have treated non-nuclear weapon states with double standard, giving a favor to certain countries at Washington's own discretion.

On the one hand, the new nuclear non-proliferation initiative announced by President Bush during his address at the National Defense University in February 2004 has an element that can trigger such complaints. For instance, he proposed the Nuclear Supplier Group (NSG) to "refuse to sell enrichment and reprocessing equipment and technologies to any state that does not already possess full-scale functioning enrichment and reprocessing plants."[84] If his proposal were implemented, South Korea would not be able to complete its nuclear fuel cycle permanently, which is contrary to the hope held by the South Korean nuclear community. Although the ROK voluntarily chose not to possess enrichment and reprocessing facilities in a way to induce North Korea to give up nuclear weapon programs, it wishes to get access to such technologies if and when the North Korean nuclear crisis is resolved and nuclear suspicions are stripped from the Korean peninsula. Moreover, to the South Koreans, President Bush's initiative can be viewed as reflecting U.S. discriminatory attitudes against South Korea in favor of Japan since Japan is the country that already possesses full-scale

---

[84] "Bush's speech on the spread of nuclear weapons," *New York Times*, February 11, 2004.

enrichment and reprocessing plants. It should be reminded that this sense of discrimination is a forceful element that fosters and sustains anti-American sentiments in South Korean society.

On the other hand, the Untied States is becoming increasingly less willing to put pressures against Pakistan and India that have become nuclear weapon states since 1998. Pakistan is a strategic partner of the United States in conducting war against terror and the Musharraf regime has established an intimate relationship with the Bush administration since the 9/11 terror. In this course, Pakistan also cooperated with the United States and the IAEA to crack down the nuclear smuggling network run by the Abdul Qadeer Khan, the developer of Pakistan's nuclear bomb. Then, the national security advisor Condoleezza Rice praised the Pakistani authorities as: "because of Pakistan's cooperation, because of Pakistan's action based on information that they've been receiving from a number of sources,..., we really now have a chance to wrap up this group [the Khan network]. And that's the most important thing."[85] However, Pakistani cooperation to curb a nuclear smuggling network organized by its citizens cannot give an acquittal to Pakistani nuclear weapon development that will remain a serious impediment to regional stability and international peace.

In case of India, the Bush administration agreed to help India to realize its goals of promoting nuclear power and achieving energy security—a commitment to provide India with a variety of civilian nuclear components from nuclear reactor to related

---

[85] David Sanger, "Bush proposes fuel ban to end spread of A-bombs," *New York Times*, February 11, 2004.

materials and technology.[86] This agreement is contrary to the long-held non-proliferation principle that countries refusing to sign the NPT should be denied civilian nuclear assistance. In order to give India civilian nuclear assistance, the NSG guidelines as well as the U.S. domestic laws should be amended. There are many criticisms on this deal even within the United States. For example, a non-proliferation expert observed that "this is a stunning example of the Bush administration's policy of exceptionalism for friends at the cost of a consistent and effective attack on the dangers of nuclear weapons."[87] In addition, concerns exist that there are many unanswered questions about implementing the U.S.-India deal.[88] Indeed, the U.S. strategic interest to counter China's power in Asia by providing such technology to India has a danger of damaging "one of our country's most strategic, effective and 'realistic' agreements: the Non-Proliferation Treaty."[89]

Export Control Based on a Principle of Disclosure

The diffusion of civil and military technologies and the proliferation of dual-use items are indeed a new reality faced by the international export control community since the end of the Cold War. The traditional policy of controlling proliferation by denying access to materials and information has been undermined in this new reality. In the post-Cold War era, national choice—not

---

[86] *Joint Statement Between President George W. Bush and Prime Minister Manmohan Singh,* Office of the Press Secretary, the White House, July 18, 2005, htttp://www.whitehouse. gov/news/releases/2005/07/ 20050718-6.html.

[87] Dana Milbank and Dafna Linzer, "U.S., India may share nuclear technology," *Washington Post,* July 19, 2005, p. A01.

[88] "Questions linger as Bush pushes India nuclear deal," *Reuters,* August 1, 2005.

[89] Lawrence Korb and Peter Ogden, "A bad deal with India," *Washington Post,* August 3, 2005, p. A19.

technical access—has become the decisive factor in many areas of proliferation.[90] Current trends in the proliferation of dangerous technologies compel a decisive shift in policy. One element of the policy shift should be a change in the principal mechanisms of control from denial of access to technology to cooperatively induced restraint.

To simultaneously achieve the two objectives—effective export control on the one hand and sustainable economic development on the other, a control regime in the future will have to focus on ways to restrict the application of technology, rather than on the increasingly futile effort to choke off supply.[91] It will also have to provide, from the production stage, necessary information about the concerned materials and items to the export control authorities based on the principle of disclosure.[92] The key is to develop a structure that allows relatively free trade of dual-use items, while at the same time, ensuring that these items are used only in civilian applications. For the purpose of this, the importance of transparency in supplying countries' export control policy and recipient countries' non-proliferation policy is emphasized in the following:[93]

> To be successful, policymakers must be able to reduce the risk of failure in the market for dual-use items by ensuring a degree of transparency and disclosure that will allow sufficient monitoring of the activities of private sector actors. In addition, and ultimately probably more important, a high degree of transparency will reduce the incentive to engage in illegal activities, thus reducing the need for elaborate and often expensive control efforts in the first place.

---

[90] Janne Nolan, *et al.*, "The imperatives for cooperation," p. 21.

[91] Ibid., p. 25.

[92] Wolfgang Reinicke, "Cooperative security and the political economy of non-proliferation," p. 179.

[93] Wolfgang Reinicke, *Global Public Policy: Governing without Government?*, p. 186.

There are direct and indirect export controls in the manner of achieving this objective. The former applies to first-hand players in export business such as manufacturer and distributor of dual-use items. The latter refers to second-hand players involved in some capacity in the production, distribution and financing of such items and those who get access to relevant information.

Two major elements of direct export control are to establish a solid import-export data base and to utilize horizontal subsidiarity. An example of a solid data base is an electronic data gathering system called KOBRA introduced by the German government in 1989. KOBRA centralizes, in a single database, all documents filed with German customs and with the licensing office, allowing for quick comparing, checking, disseminating, and exchanging important information regarding suspicious behavior on both the supply and the demand sides. Since then, customs authorities in many countries have started to use more sophisticated large-scale electronic systems with the aim of achieving paperless export and import, for example, Australia's EXIT, Great Britain's CHIEF, and the United States' AES.

Horizontal subsidiarity encourages mutual learning among the government branches and private interest groups, which is a significant precondition for achieving greater cooperation among them. For example, the idea of a know-your-customer (KYC) policy for dual-use trade is an option to expand horizontal subsidiarity. Sales and marketing agents of a supplying company and employees of related financial institutions can be effective export control checking points because long-established client relationships lead to insider information about the purpose of a specific order and familiarity with the historical pattern of the customers' purchases makes it possible to judge whether the order is in any way suspicious.[94]

For effective indirect control, policy makers must get helping hands from other market participants to provide information, thus creating a network of data from multiple sources. Financial institutions will be central to this effort because export/import behaviors will ultimately have to end in the financial transactions. In this regard, the KYC policy also applies to the indirect export control whose necessity was articulated as follows:[95]

> Financial institutions themselves need to expand their KYC policies to include the proliferation threat as well, by tracking certain types of data that have proved valuable in detecting illegal dual-use trade. These include, for example, financial transfers from a proscribed destination, especially when the funds go to companies or other end users that have a history of violating export control laws…., Similarly, when preparing a loan or underwriting a stock or bond issue, a financial institution has access to detailed information about its intended purposes. The bank could establish specific guidelines for providing credit or underwriting bonds or equity for projects, either in a particular country or of a specific nature. Evaluation of this information could reveal a project's proliferation potential at a very early stage.

---

[94] Ibid., p. 194.
[95] Ibid., p. 196.

Chapter Four

Greater Nuclear Transparency and the ROK

Despite efforts to demonstrate its intention to make use of nuclear energy for peaceful purposes only, concerns of the international community about South Korea's possible nuclear proliferation have not faded away. Those concerns are obviously misrepresenting the ROK government's determination to devote itself to peaceful uses of nuclear energy for the welfare of Korean people and hindering its research and development activities for that purpose. South Korea's geopolitical situations, surrounded by three nuclear powers and one potential nuclear power and in particular, military standoff *vis-a-vis* North Korea may be the major factor to spin such biased views. For example, the U.S. Department of Energy noted that North and South Korea interact dangerously with painful energy vulnerabilities, storage problems, and *political-military incentives to at least seriously consider nuclear weapons* [emphasis added].[96] Another factor that might

---

[96] U.S. Department of Energy, "Policy forum: energy futures," *Washington Quarterly*, Autumn 1996, p. 94.

have created international suspicions of nuclear intention of the ROK is the lack of sufficiency in the transparency of its non-nuclear policy.

Enhancing transparency of non-nuclear policy is meant to disclose sufficient information in the course of planning and implementation of the policy and thereby, to demonstrate peaceful intentions and activities of the ROK government in making use of nuclear energy. It will have the effect of eliminating international suspicions of South Korea's possible nuclear weapons development and of reinforcing its commitment to the denuclearization of the Korean peninsula. Enhancing transparency of non-nuclear policy is vital for maximizing national interests of the ROK in several ways.

Conforming to the International Non-Proliferation Trends

As discussed previously, the more the world becomes linked globally, the wider the dimension of common interests expands. One major common interest is to dissuade the risks and dangers of proliferating weapons of mass destruction. In this respect, it has been notably a clear trend of the international society that norms and institutions have been strengthened to curb various attempts in many areas of WMD proliferation. This trend is well illustrated by the fact that the Non-Proliferation Treaty is the second biggest international treaty next to the United Nations membership. As of 2005, 191 countries signed up the United Nations while 189 nations did the NPT.

Under the circumstances, to make its non-nuclear policy full of ambiguities is to violate a most widely accepted international norm of non-proliferation. Such a move would be indeed morally irresponsible and politically reckless. In practice, the ROK is expected to receive enormous disadvantages if it does not eliminate nuclear suspicions. For example, an obscure non-nuclear policy could harm the credibility of South Korea's national policies as a whole. In a tightly interwoven international society in the 21st century, nuclear discredit would isolate South Korea, diminish its diplomatic capabilities, thus bringing about many difficulties in key issue areas.

Increasing Korean People's Credibility and Image

The DPRK is the only country in the world that had violated the NPT twice and finally broke away from the treaty. It is also one of a few countries yet to sign on the Chemical Weapons Convention (CWC). North Korea's reckless behavior against international norms and rules has hardened bad images of the country as an unreliable and unpredictable rogue state trying to do all sorts of messy things. It further fixated the North Korean leadership as a dictatorial regime obsessed to cling to power at all costs while taking Korean people hostages. It also gives added credits to the Bush administration's rigid perceptions and approaches toward the Kim Jong Il regime.

As the DPRK makes troubles for the world, it becomes a burdensome duty for the ROK to recover the Korean people's credibility and image in the world. Because the two sides will become integrated as one nation, maintaining good image and credibility of Korean people in the world community is an important matter for the better future of Korea.

In this regard, the North Korean regime's ill-natured behaviors related with its nuclear programs must be clearly pointed out.

Looking back to the history of North Korea's nuclear development, "deception" and "persistence" may be the two words that most succinctly describe the North Korean regime's psychology and strategy on nuclear weapons. Throughout its history, the North's nuclear weapon development program has been disguised by the Pyongyang regime's peaceful rhetoric of having no intention to go nuclear. North Korean authorities, of course, stubbornly exerted themselves in furtive efforts to acquire nuclear weapons at the back door. Under the banner of "having neither intention nor capability to develop nuclear weapons," guided by the late President Kim Il Sung, this pattern of rhetorical deception on the one hand and persistent obsession about nuclear weapons on the other had continued until April 2003 when North Korea finally revealed that they had nuclear weapons.[97]

There have been several examples manifesting North Korea's duality and dishonesty. First, by signing the Joint Denuclearization Declaration with South Korea in 1991, North Korea promised not to possess reprocessing or enrichment facilities. But the IAEA inspection that was carried out just six months later found that the North had already constructed and operated a large-scale reprocessing facility—what they called a radiochemical laboratory. Indeed, the Joint Declaration was a stillborn child from the beginning. The amount of weapon-grade plutonium produced by North Korea before the IAEA inspection

---

[97] It was during the conversation with the editor-in-chief of NHK in October 1977 that North Korean President Kim Il Sung first publicly expressed his intention not to develop nuclear weapons. At an interview with the President of Iwanami Shoten on September 26, 1991, he declared to have neither intention nor capability to develop nuclear weapons. At a luncheon with the South Korean delegation for the South-North High-Level Talks on February 20, 1992, Kim Il Sung stated that "we do not intend to have a nuclear confrontation with neighboring big powers and in addition, it is unimaginable to develop nuclear weapons that can wipe out Korean people." *Rodong Sinmun*, February 21, 1992.

in 1992 is estimated to be around 10-14kg.[98]

Second, the Pakistani government's investigation of Dr. Abdul Qadeer Khan and subsequent revelation of his nuclear smuggling network in early 2004 showed that there had been a significant level of nuclear cooperation between North Korea and Pakistan. During the last decade, technologies, equipments and materials related to uranium enrichment had flown from Pakistan into North Korea. Abdul Qadeer Khan, the father of Pakistan's nuclear weapon, visited North Korea more than a dozen times. Under investigation of his nuclear smuggling network, Dr. Khan told Pakistani investigators that he was engaged with North Korea on the sale of HEU equipment and saw three nuclear devices while visiting Pyongyang in late 1990s. [99] The international intelligence community has also begun to reveal some of the considerable material on the DPRK HEU program. For instance, according to the U.S. National Intelligence Estimate of June 2002, the CIA understood that Pakistan had shared with the DPRK high-speed centrifuge technology, information on construction of a uranium-triggered nuclear device, and test data of such a weapon.[100] This is a clear violation of the Joint Declaration, the Geneva Agreed Framework, and the Non-Proliferation Treaty (NPT).

Third, pointing to the United States as the major source of tension on the Korean peninsula and of the breakdown of the Agreed Framework, North Korea finally withdrew from the NPT and became the first such nation in the history of non-proliferation. At the NPT withdrawal statement issued in January 2003, the DPRK

---

[98] *The Defense White Paper* (Seoul: The Ministry of National Defense, 2004), p. 39.
[99] David Sanger, "Pakistani tells of North Korean nuclear devices," *New York Times*, April 13, 2004.
[100] Seymour Hersh, "The Cold Test," *New Yorker*, January 27, 2003.

government reasserted itself that it did not have any intention to go nuclear and invited the United States to verify their statement. About three months later, the government statement was nullified at the Beijing three-party talks when the DPRK representative Lee Gun informed to the U.S. representative James Kelly that North Korea already had nuclear weapons.[101] Mr. Lee's remark was the first case where a high-level North Korean authority revealed that Pyongyang possessed nuclear weapons. Since June 2003, North Koreans set to speak out that they have a "nuclear deterrent force."[102]

In short, what the North Korean regime has shown to the international society as regards to its nuclear ambition is indeed a historical masterpiece of ill-natured deception and unyielding persistence. Threats posed by North Koreans will be brought to an end only when such persistent deception no longer serves as a guiding principle of their strategic thinking and policy-making behaviors.

In this regard, enhancing transparency of the ROK non-nuclear policy is important for the Korean people to acquire credibility and respect as a responsible member of the international community, and to maintain the esteem and dignity. In fact, it is a

---

[101] Foreign Minister Paik Nam Soon and Vice Foreign Minister Kim Gye Gwan reconfirmed North Korea's possession of nuclear weapons when they met a U.S. Congress delegation led by Representative Curt Weldon in late May 2003. *Dong-a Ilbo*, June 3, 2003.

[102] A commentary of the *Korean Central News Agency* argued that "if the U.S. keeps threatening the DPRK with nuclear weapons instead of abandoning its hostile policy toward Pyongyang, the DPRK will have no option but to build up *a nuclear deterrent force* [emphasis added]." *Korean Central News Agency*, June 9, 2003. Before this commentary, on June 6, spokesman for the DPRK Foreign Ministry said that "as far as the issue of a nuclear deterrent force is concerned, the DPRK has the same legal status as the United States and other states possessing nuclear deterrent forces which are not bound to any international law." *Rodong Sinmun*, June 7, 2003.

matter of national image and pride. The ROK non-nuclear policy is a countering force that will help strip the international community of bad images and wrong perceptions of Korean people.

Lacking Technical Capabilities to Go Nuclear

Even if the ROK has an intention of going nuclear, it is lacking necessary technical capabilities to do so. In the 1970s, then President Park Chung Hee had attempted to run a nuclear weapon program as a bid to counter the United States' withdrawal of its forces from South Korea. The technical infrastructure was entirely dissolved right after his death in 1979. In addition, the ROK government has adhered to the Denuclearization Declaration and according to this, there exist no programs related with reprocessing or enrichment in South Korea. It is virtually impossible for the ROK to operate an indigenous nuclear weapon program without being detected by international supervisions. It has zero possibility as well that a nuclear weapon country including North Korea would help South Korea develop nukes. Under the circumstances, an ambiguous non-nuclear policy would only bring about political suspicions of proliferation.

Lessening Barriers to the Peaceful Uses of Nuclear Energy

Nuclear power is the key energy source in the ROK, as shown by its current reliance on nuclear energy for more than 38 percent of its electricity demand. This trend will continue in the foreseeable future. Unless alternative energy resources are found, dependence on nuclear energy will be growing. So in terms of energy security, peaceful uses of nuclear power have become a critical element of South Korea's energy policy. The problem with nuclear suspicions is that it would cause visible or invisible adverse effects that stand in the way of the ROK nuclear industry's R&D activities.

On the other hand, the more transparent its non-nuclear policy is, the less suspicious the international community would become of the ROK non-nuclear policy. And consequently, the barriers standing in the peaceful uses of nuclear energy would become lessening. Japan is one of the best examples in this respect. According to a Japanese expert, Japan would not go nuclear despite North Korea's nuclear threat. The main reason is that if Japan breaks away from the NPT, Japan has to expect that a lot of sanctions will be imposed upon itself. One of them must be termination of nuclear fuel supplies from abroad, which will cause hazardous impacts on Japan's energy security and nuclear industry.[103]

Contributing to the Peaceful Resolution of North Korea Nuclear Crisis

North Korea's nuclear problem of today is different in many aspects from that of ten years ago. There are at least four differences noted.

Firstly, North Korea's American counterpart is different. Compared to the Clinton administration, the Bush administration has very different perceptions on the leadership of North Korea and takes fundamentally different approaches toward the DPRK. Such differences are highlighted in demanding higher and more

---

[103] Author's conversation with Fumihiko Yoshida, editorial writer of the *Asahi Shimbun*, June 13, 2005.

rigorous level of transparency and verification. Distressed with providing incentives to rogue states for scraping their misdemeanor that should not have occurred in the first place, the U.S. Republican party has been a vocal critic of the Clinton administration's North Korea policy, the tone of which is inherited in the Bush administration. For instance, Henry Hyde elaborated a hard-nosed Republican position on the DPRK, saying that verification is the key to dealing with North Korea since the DPRK's demonstrated willingness to embrace adequate verification measures is "a signal of a genuine break with the past and a commitment to future cooperation."[104]

Secondly, there have been dramatic changes in the international security environment since the 9/11 terror accident. Since 9/11, it has been regarded as a part of a war against terrorism to bar rogue regimes and terrorist groups from developing WMD. International understanding and cooperation against WMD proliferation has never been as strong as today. Whoever the target is, multilaterally coordinated efforts, often being coercive, will be justified with full support of the global community. North Korea is no exception in this context. Patrons of North Korea— China and Russia—having their own war against terrorism, will not be able to make valid objections to the pressing approaches against the North when needed.

Thirdly, today's nuclear problem is a reality whose existence was confirmed by North Korea. The DPRK declared in February 2005 that it has made nuclear weapons and would further bolster its nuclear arsenal. However, the nuclear problem in the 1990s was

---

[104] *Henry Hyde's Speech at the American Enterprise Institute* in Washington, D.C. on March 13, 2001. See http://www.nautilus.org/napsnet/dr/index.html#item2.

and has remained as a suspicion due to Pyongyang's persistent denial. This means that North Korea cannot be justified at this time in refusing international demand to reveal all necessary information, to dismantle relevant facilities, and to fully cooperate with the IAEA to have thorough inspections.

Finally, North Korean declaration of their making nuclear weapons is a full proof that it has violated four major international agreements: the NPT, the IAEA Safeguard Agreement, the Joint Denuclearization Declaration and the Agreed Framework. Making little of international obligations it assumed, the North Korean regime is indeed a renegade leadership from the world. This gives added credits to the Bush administration's rigid perceptions and approaches toward North Korea.

These differences between the early 1990s and today illustrate stark difficulties involved in the present crisis. The ROK government's solid non-nuclear policy will help resolve the problem in three ways. Firstly, it will not provide North Korea with any excuse to either justify its nuclear weapons or delay the negotiating process as it did in 2004 by linking the 2004 incident of South Korea with its participation of the six-party talks. Secondly, it will set a role model that must be followed by the North Korean regime, thus producing political and diplomatic pressures upon Pyongyang. Thirdly, it can present the ROK government with a better opportunity to play a leading role in the course of resolving the nuclear crisis. This opportunity comes from both moral and practical strength in that the ROK—as a nation whose security is most threatened by North Korean nuclear weapons—holds on to the non-nuclear policy, shattering any hint of violating the international non-proliferation regimes.

Preventing Arms Race in Northeast Asia

Lack of transparency in the ROK non-nuclear policy and resultant nuclear suspicions will keep neighboring states in constant nervousness. As was mentioned before, geopolitical and strategic circumstances surrounding South Korea have tempted the international community as well as regional countries to be suspicious of the ROK government's nuclear intentions. Undoubtedly, not transparent non-nuclear policy full of ambiguities will make nuclear suspicions deeper and wider. This might induce unnecessary tension and could cause an arms race in the region—a boomerang, which is obviously not what the ROK government and people would like to see. A transparent non-nuclear policy upheld by the ROK government will foster positive environments for preventing military tension and arms race in Northeast Asia.

Fostering Auspicious Atmosphere for Korean Unification

North Korea's nuclear showdown with the world presents two important policy implications for Korean unification and non-nuclear policy of South Korea.

Firstly, since North Korea's bad images in the international community are worsened, South Korea will bear much more burdens in the future process of unification. Unless North Korea grows mature enough to be a responsible member of the international society, unification of the two Koreas cannot attain international support and assistance, which is an essential component of unification. Therefore, South Korea, with the helping hand of the world, should put more efforts to bring about real and constructive changes in North Korea and to keep the

North Korean regime in a peaceful domain.

Secondly, Korean unification will neither be feasible nor welcomed unless the international community firmly believes that unification does not disturb regional stability and peace. In this context, here is a growing importance to eliminate international suspicions over the two Koreas' ambitions to possess nuclear weapons. In this regard, Seoul is in a far better position than Pyongyang. But recent public attitudes in South Korea toward North Korea's nuclear problem, for example, emotional understanding of Pyongyang's nuclear weapon program, pointing Washington as a source of the problem and putting national cooperation among South and North Korea ahead of international coordination, could taint the integrity of South Korea's non-nuclear policy. South Korea should exert more efforts to educate general public about why sticking to the policy is important for the Korean nation's interests.

That is, believing that a unified Korea would go nuclear, neighboring countries understandably would make every effort to stand in the way of Korean unification. Unless South and North Korea make sure that they are non-nuclear and will remain so in the future, they cannot expect the external support and assistance that will be essential in the unification process. It should be remembered that West Germany's strong advocacy that unified Germany would not pursue weapons of mass destruction facilitated German unification by allaying the security concerns of neighboring states as well as the four key countries.[105] In a recent national security report of the Untied States, the concern of

---

[105] Karl-Heinz Kamp, "Germany and the future of nuclear weapons in Europe," *Security Dialogue*, Vol. 26, No. 3, 1995, pp. 277-292.

unified Korea's nuclear possession also led to an argument that the U.S. forces should remain in Korea after unification in order to ensure a non-nuclear Korean peninsula.[106] For Koreans, a nuclear weapon option is a useless "card," if it was ever thought to be so. It should be readily discarded for the more sacred and desperate goal of national unification.

---

[106] The United States Commission on National Security/21st Century, *Seeking A National Strategy: A Concert For Preserving Security And Promoting Freedom*, April 15, 2000.

Chapter Five

The ROK Position on Nuclear Transparency

Compared with most developed countries, a policy-making culture in the ROK has been incomplete or deficient in many ways. This reality hints that the ROK government has not achieved two essential components of public policy-making: (1) to devise and establish policies through a scientific and reasonable process, and (2) to implement the policies in an effective and efficient way based on persuasive logics and convincing explanations. In particular, it is typical to note in South Korea that many policy makers have a tendency not to disclose what they know about a specific policy—that is, reluctance to policy transparency. Policy makers often try to avoid from revealing details related with a policy under the name of "protecting sensitive information."

Such anti-transparency tendency could cause several problems. Firstly, since the policy-making process becomes ambiguous, it is difficult to hold someone accountable when a policy turns out to be wrong or to harbor serious mistakes. Consequently, this will create an easy-going atmosphere in the bureaucracy. Unless the process of policy-making becomes transparent and responsibilities of individuals are clarified, it will be difficult to shy away from insolvency and obscurities in policy-making.

Secondly, expertise necessary for making a good-quality policy in an issue area cannot be guaranteed. In a secretive and loose policy-making culture, inexperienced individuals are easy to get heavily involved in the policy-making process. In addition, any problem or mistake can be easily covered up, promoting amateurism in the policy-making community.

Thirdly, since a clear distinction is not established between what to be disclosed and what not to be, policy-makers tend to miss a sense of obligation to protect sensitive information that is an essential virtue of a responsible expert. As a result, an ironic situation occurs when a really important issue that should be kept secret is revealed without mal-intention. In many occasions, far too much effort is being wasted protecting not really secrets, which allows vital secrets to slip through. Since so much is classified, it is often impossible for people with security clearances to know what is derived from a spy satellite and what is plucked out of a newspaper. This phenomenon is called "the cult of classification," in which information both rare and commonplace is safeguarded with equal zeal.[107]

It is essential for short-term as well as long-term interests of South Korea to invent sound positions and attitudes on enhancing transparency of the non-nuclear policy. For the purpose of

---

[107] Stratfor.com's Weekly Global Intelligence Update, September 18, 2000, http://www.stratfor.com.

achieving this objective, the following measures are presented as guidelines for formulating a better position on nuclear transparency and increasing the level of transparency of the ROK non-nuclear policy.

Establishing Positive Stands on Nuclear Transparency

A first step should be to increase the understanding of policy-making and scientific communities about transparency—i.e., what transparency means, what positive roles it is to play, and what negative effects it could create, etc. Only such a thorough understanding about transparency and international trends for promoting transparency can prevent reluctance or misperception in relation to transparency. It is also a precondition for the ROK government to establish clear and reasonable goals by accepting the international standards of transparency in the nuclear field. Succinct understanding and clear goals are the basis upon which directions and means for an effective non-nuclear policy are established.

Not only for non-proliferation, nuclear transparency is also necessary for the safety of operating nuclear power plants. That is, transparency in the nuclear field is needed to cultivate "safety culture" for secure and sustainable uses of nuclear energy. In terms of safety culture, the ROK has recorded good marks by maintaining high level of transparency in accordance with international safety standards. Thus, policy guidelines will focus on the non-proliferation aspect.

The question is what degree of transparency should be needed and how much information should be disclosed to demonstrate that the ROK has no intention to divert its nuclear research and development activities into military purposes. Here comes an issue of compromising sensitive commercial information and academic achievements. While transparency *per se* is important, the international community is well aware of significance of protecting sensitive information for commercial interests and property rights. This awareness was illustrated in the cautious approaches used in the Chemical Weapons Convention. Of course, nuclear weapon states will try to gather as much information as possible about non-nuclear weapon states' nuclear activities and to minimize uncertainties as much as they can. In this respect, a pitfall of transparency should be noted that an element of psychology and perception is placed in the debate of transparency. This means that even a complete disclosure of whatever information possible may not be enough to assure a perfect transparency, leaving some uncertainties beyond physical dimension. In this perspective, a reasonable role of transparency must be to minimize but not to completely eliminate uncertainties.

There are two criteria to judge whether nuclear proliferation occurs in a country. The first criterion is the country's intention to develop a certain nuclear R&D program in the first place. For instance, if a country pursues a sensitive program such as running a reprocessing plant without having economic or scientific justifications, a strong suspicion will be raised against that country. Such a suspicion would occur after a sensitive program or related activities are initiated, and thus, it is defined as *dependent suspicion*. In case a country is regarded as unreliable and dangerous by the international society, dependent suspicion and subsequent caution of the international community will be multiplied. Proper examples are North Korea, Iran and Iraq. On the other hand, if a country is accepted as a favorable member of the international community with good image and credibility, dependent suspicion, if any, will be either minimized or neglected. Good examples are Japan, the Netherlands, and Germany.

The second criterion is the country's technical capability. Even if a country got rid of dependent suspicion of its intention, the constant worries could remain that someday its technical capability may be diverted for military purposes. It is defined as *inherent concern*. For instance, despite Japan's nation-wide efforts to manifest its determination not to develop nuclear weapons and to enhance transparency of its non-nuclear policy, an inherent concern has been kept sprung out intermittently, which is derived from its advances in broad scientific capabilities as well as in nuclear technologies.

South Korea is tied in a difficult position. On the one hand, facing geopolitical and strategic situations as described earlier, dependent suspicion will be raised whenever a sensitive nuclear R&D program is initiated. On the other hand, the ROK nuclear community has reached a certain level of technical capability though not fully advanced as Japan and other western European countries and inherent concern will arise due to the technical capability. Thus, if the ROK is found to have a sensitive R&D program as was revealed in 2004, not surprisingly, dependent suspicion and inherent concern is likely to be fermented.

It is, in particular, noted that dependent suspicions and inherent concerns are partly driven from how a target country is perceived. Such an external perception is forged by a spectrum of elements from the various disciplines—ranging from politics, security, foreign affairs, history, culture, economy, science and technology including nuclear field. Therefore, international suspicions and concerns of a country can be regarded as an outcome of its foreign relations, domestic politics, national power, scientific prowess, and national image at a particular point of time. The 2004 incident had better be put into this perspective in order for the ROK government not to repeat any further mistake.

The concept of transparency needs to be understood in a broader context than just a narrowly focused give-and-take relationship between audiences and actors.[108] Any country can play a dual role both as an audience and an actor, depending on the situations. For instance, the ROK government is an actor *vis-à-vis* the United States while it is an audience in relation with North Korea. This implies that when the ROK government formulates its stance on transparency, it should not focus only on the actor role, which will harbor the danger of South Korea becoming too defensive to the idea of greater transparency. Defensive arguments focusing on the audiences such as the United States, the IAEA, and others would not effectively represent national interests of the ROK. Those arguments could be out of perspective because they can be contradictory to the ROK logics to demand North Korea for disclosing greater information about its nuclear weapons program.

Therefore, when the ROK government develops its stance on transparency, it should consider dual roles both as an audience and an actor and take comprehensive issues surrounding transparency into account. Having this in mind, this study offers the following stance of transparency as a consensual understanding

---

[108] An audience refers to a party to demand disclosure of information and the actor is a party to be asked to increase transparency. For example, in the IAEA-ROK relationship, the IAEA is the audience and the ROK government is the actor. In general, the international society incorporating major technology-supplying countries and international institutions on the one hand and the domestic public opinion on the other hand are two pillars of audiences that are to be faced by a country.

among the ROK government, scientific communities and general public:

> Transparency is a comprehensive concept covering various national policy-making areas from politics, diplomacy, economy, security, environment, society, science, history, culture, etc. Demand on greater transparency in the nuclear field occurs in the relationship between an actor pursing a specific nuclear program and audiences such as technology-supplying countries, international institutions or domestic opinions having concerns about the program. From the psychological point of view, the issue of "how much transparency is enough" is affected by subjective judgments driven by qualitative elements like an actor nation's image, credibility, and its relationship with audiences. From the technology point of view, enhancing transparency is the process of reducing uncertainties regarding an actor's intentions and activities by the process of audiences' information collection and an actor's information disclosure.

The ROK government can establish objectives and intended outcomes of greater transparency in non-proliferation and safety areas, respectively, depending on the two pillars of audiences.

For Domestic Opinions in Non-Proliferation Area
The objective is to assure general public that the ROK government's peaceful uses of nuclear energy are in no violation of its non-nuclear policy. The intended outcomes would be to increase the level of public trust to the ROK government's policy of using nuclear energy solely for peaceful purposes and to cultivate South Korean people's minds supporting non-proliferation and peaceful uses of nuclear energy.

For Domestic Opinions in Safety Area
The objective is to guarantee South Korean public that nuclear energy is safe, environment-friendly, and is contributing to the

welfare of South Koreans by making use of scientific data and technical evidences. The intended outcomes would be to reduce reluctance to nuclear energy from South Korean people and induce their support for it, and in consequence, to help acquire necessary real estates for building new nuclear power plants and storing nuclear wastes, and to foster auspicious circumstances for further development of nuclear industry.

For International Society in Non-Proliferation Area
The objective is to assure the international society that peaceful uses of nuclear energy in the ROK are in full compliance with the international non-proliferation regimes. The primarily intended outcome would be to root out misunderstanding, misjudgment, or suspicions that the ROK government may be pursuing to develop nuclear weapons. Other intended outcomes would be to increase national credibility and image of the ROK, to contribute to strengthening non-proliferation regimes, to help reduce unnecessary tension and mistrust and establish more friendly relations among the countries in Asia, and to induce the nuclear-supplying countries to increase greater technical cooperation with the ROK without hesitation.

For International Society in Safety Area
Two objectives are to comply with international norms and rules about safety and to contribute to developing more advanced safety culture and relevant technologies for that purpose. The intended outcomes would be to reinforce a strong status as a nation with advanced commercial nuclear infrastructure in the international non-proliferation community, to establish a constructive national image by becoming a role model in the field of nuclear safety, and to foster a favorable international environment to promote further development of nuclear industry

in the ROK.

In conclusion, enhanced transparency in non-proliferation and safety will help the ROK government establish a sound and solid foundation that makes stable and sustainable uses of nuclear energy possible. International support and credibility acquired by increasing transparency will make it possible to strengthen both domestic and external bases for promoting peaceful uses of nuclear energy, which is a shortcut for booming nuclear industry. In the wake of the 2004 incident, many people in the South Korean nuclear research community came to voice concern that the ROK should be vigilant in keeping up full nuclear transparency in order to dispel any suspicions in nuclear activities,[109] which means that South Korean nuclear community is moving for the right direction. The task ahead should be to develop and implement a series of detail measures to enhance transparency in the ROK non-nuclear policy.[110]

Understanding the Complexities Involved in Transparency

The policy-making and nuclear communities in the ROK should have a firm grasp on a reality that enhancing policy transparency in a sensitive area like nuclear energy is a complex issue involving multiple considerations and having influences on other policy-making areas. As described before, due to globalization, the international system has been getting integral and member

---

[109] Hee-seog Kwon, Director of Non-proliferation and Disarmament, Ministry of Foreign Affairs and Disarmament, "Lessons and perspective of nuclear transparency in Korea," a paper presented at the Seminar on *Nuclear Energy Non-proliferation in East Asia*, organized by Korean Nuclear Society and Sandia National Laboratories, on August 24-26, 2005, in Seoul, South Korea.
[110] Interview with Dr. Jungmin Kang, Research Fellow at the Center for Nuclear Policy, Seoul National University, September 12, 2005.

nations of the system have become mutually dependent. In addition, major policy issue areas are being linked in one way or another, and various policy-related communities within and without a nation have been engaged through visible and invisible networking.

In this respect, a dual-use industry such as nuclear energy stretches its relationship to many areas like domestic politics, foreign affairs, economy, security, environment, etc. So transparency in such an industry has to take into account various problems arising from this complexity. Particularly, politics of transparency should be noted here. When an actor is alleged to be lacking nuclear transparency, the audience may have raised such allegation due to a reason or aim that was not expected by the actor in the nuclear domain. For example, if a particular audience claims that the ROK government should increase transparency of its non-nuclear policy in general or a particular R&D program, it could be more than just a demand on disclosing some information or clarifying technical details of the program. An audience's demand on transparency in a particular moment of time is possible to be a diversionary tactic beyond technical dimension. It could be thrown out to put pressures on the ROK government in broader diplomatic, security, and economic contexts or to draw concessions from the ROK government in other contending issue areas.

Demand for transparency also represents domestic politics in an actor and its sensitivity perceived by an audience. For example, if anti-American sentiments grow and nationalism intensifies in South Korea, inherent concerns and dependent suspicions of the ROK nuclear R&D activities will be multiplied, which will lead to request for greater than usual transparency. In this context, it is noted that the year 2004 was the period when anti-American sentiments and nationalism were unusually high in South Korean society.

Proper understanding of the complexities involved in transparency will lead the ROK government to take account of a broad spectrum of factors when an allegation is raised regarding transparency of its non-nuclear policy. Only if all these factors are properly paid attention to in a comprehensive way, it would be feasible for South Korea to shape a credible, durable and integral non-nuclear policy. A durable non-nuclear policy would be such that would not be altered easily if some of the underlying factors change.

Addressing the Psychological Aspect of Transparency

Transparency is, in large part, a subjective matter of perception that an audience harbors about the intentions and activities of an actor. There exist many factors in various fields such as history, culture, diplomatic relations, economic exchanges, and security that would affect the perception of an audience. This indicates that transparency-related problems could be prevented or its seriousness ameliorated if the ROK government's non-nuclear policy—from its formation to implementation—addresses the psychological aspect of transparency. The following points should be noted in this regard.

Firstly, it would be wise for the ROK policy-making and nuclear communities to avoid any remarks or behaviors that could trigger suspicions or misunderstandings on the part of an audience. Exclusive thoughts, nationalistic opinions, and peculiar behaviors contrary to international norms and standards should be avoided not only by policy-makers, high-profile politicians, and nuclear scientists but also by public opinion leaders in South Korea. In particular, the epistemic community in relation to non-proliferation in South Korea could play a fascinating role in this regard.[111] There are several reasons. What the ROK epistemic community thinks and says is an important yardstick to forecast what the ROK government has in mind. It can have a significant influence on the government's basic thinking as well as detail action plans. And the ROK epistemic community has an extensive network with similar communities in other nations and thereby, its influence stretches over to the international community.

Secondly, it is important to forge and keep better relations with possible audiences of its nuclear R&D programs. Major nuclear industrial countries like the United States, France, Great Britain, Japan and nuclear-related international organizations such as the IAEA are counted as possible audiences. Friendly relations with them would make it possible to moderate excessive reactions from them if any transparency-related problem occurs in the ROK. Finally, diplomatic efforts of the ROK in the area of non-proliferation should be doubled with an aim of fostering auspicious national image and increasing national credibility, the details of which are discussed in the next section.

---

[111] For the general constructive roles to be played by an epistemic community, refer to Peter Haas, "Introduction: epistemic communities and international policy coordination," *International Organization*, Winter 1992, pp. 1-35; Emanuel Adler and Peter Haas, "Conclusion: epistemic communities, world order, and the creation of a reflective research program," *International Organization*, Winter 1992, pp. 367-390.

Highlighting Non-Proliferation and Transparency in Diplomacy

Looking back to the history of nuclear development in South Korea, it is regrettable that the importance of diplomacy has been rather neglected. There are two pillars that prop up peaceful uses of nuclear energy: scientific technology and diplomacy. Delicate diplomacy is as important as developing technical infrastructures that creates auspicious external environment for bolstering peaceful uses of nuclear energy and eliminates dependent suspicions and inherent concerns of the international community. Without elegant diplomacy supported by credibility, consistency and expertise on the nuclear field, international trust on a nation's non-nuclear policy can hardly be attained or sustained in the long-term.

Importantly, the ROK foreign affairs community must have a firm grasp on the fact that non-proliferation is indeed one of the most important consensuses of the international community in the 21st century. The cohesion of this consensus is becoming stronger in light of a series of terror accidents beginning from New York in September 2001, Madrid in March 2004, and London in July 2005. International regimes—consisting of principles, norms, rules, and institutions—on curbing proliferation of weapons of mass destruction worldwide have been reinforced through cooperative engagement in the post-Cold War era.[112] Modern states are bound in a tightly woven fabric of international regimes through treaties, tacit agreements, code of conducts, international

---

[112] A regime is typically defined as "set of implicit or explicit principles, norms, rules, and decision-making procedures around which actors' expectations converge in a given area of international relations." Stephen Krasner, "Structural causes and regime consequences: regimes as intervening variables," in Stephen Krasner, ed., *International Regimes* (Ithaca: Cornell University Press, 1983), p. 2.

organizations, and multilateral institutions. Integrity of the international system, interdependence of member nations and complexity of issue areas in the 21st century have made it increasingly difficult for a single powerful nation to exercise its strength exclusively and determine a course of an event unilaterally. The best example would be the quagmire experienced by the United States in Iraq by disregarding the majority voices expressed in the United Nations.

In this regard, it is noted that sovereignty of a nation state—the vindication of the state's existence as a member of the international system—needs to be redefined. According to one study, the only way most states can realize and express their sovereignty would be through participating in the various regimes of the international system and thus by complying with relevant principles, norms, rules, and regulations of institutions.[113] This is the way new sovereignty is exercised properly. In this compliance-prone external environment, the nature of foreign policy also changes as follows:[114]

> The traditional attributes of effective foreign policy in the security area—flexibility, energy, secrecy—tend to give away before the growing importance for the new sovereignty of predictability, reliability, and stability of expectations [of the international community]..., Compliance with the norms governing this environment becomes not so much a curb on the will or preferences of the state as a condition for realizing the full range of its objectives.

In relations to this, the importance of increasing transparency must be highlighted in the ROK foreign policy. The ROK foreign

---

[113] Antonia Chayes and Abram Chayes, *The New Sovereignty: Compliance with International Regulatory Agreements*, p. 27.
[114] Ibid., p. 124.

affairs community must acknowledge that transparency is a critical element in the integral, interdependent, and complex international community both as a means to exercise a nation's sovereignty properly and a tool to realize its national interests and foreign policy objectives. With this firm acknowledgement, the ROK diplomacy in non-proliferation affairs will have to be directed toward less exclusiveness, better compliance, less secrecy and more transparency.

Extending the Public Basis of Non-Nuclear Policy

A domestic obstacle to promoting peaceful uses of nuclear energy in South Korea is a groundless, emotional, but widely spread national sentiment for developing nuclear weapons. According to one survey conducted in 1999,[115] an absolute majority of South Korean public would be in favor of having nuclear weapons in the following four contingencies:

- In case North Korea possesses nuclear weapons (82.3%).
- In case the ROK-U.S. security alliance is ended (85.5%).
- In case Japan becomes a nuclear weapon state (86.9%).
- In case there is an external security threat after Korean unification (75.8%).

On the other hand, those who were against South Korea's nuclear weapons in any circumstance only remained at 15.0%.

---

[115] Norman Levin, *The Shape of Korea's Future: South Korean Attitudes Toward Unification and Long-Term Security Issues* (Santa Monica: RAND, 1999), p. 23.

Of course, most of South Korean people who responded to this survey must have been laymen who neither had enough knowledge about geopolitical and strategic situations of the ROK and stark international non-proliferation trends nor had been accustomed to analytical thoughts about pros and cons of having nuclear weapons. That is, the figures merely reflected emotional sentiments of Korean people at most. It should be acknowledged that the public survey is not a complete means for understanding the general opinions and reactions of South Korea.

However imperfect the survey was, notably, the high profile figures themselves produced many lessons that deserved serious attention of the ROK policy-making and nuclear communities. The importance of public opinion in a democratic country cannot be emphasized too much. Public opinion is a barometer through which the outsiders could figure out what is going on now and what will happen in the near future in that country. This means that the result of a public opinion survey—especially conducted by a prominent foreign institute—could have significant repercussions on the viewers in the international community, especially those who have paid close attention to the ROK nuclear activities. In this respect, it should be taken by the ROK government to be a serious issue that pro-nuclear weapon opinions in major contingencies were above 75%. In the similar context, it is also worried that after North Korea declared that it had made nuclear weapons, an opinion has emerged in South Korea that favors the ROK to make nuclear weapons to counter North Korean nuclear threat.

In August 2005, a public opinion survey was jointly conducted at the Donga Ilbo and the Asahi Shimbun—two prominent newspapers of Korea and Japan.[116] According to this survey, Koreans who gave positive answers to nuclear-armed South Korea reached 52%, which is ostensibly higher than the Japan's 10%. Those who are in their thirties and forties showed stronger aspiration for South Korea having nuclear weapons than other age groupings. Korean respondents who agreed on North Korea's having nuclear weapons also got to 41%.

All these societal phenomena shown in the ROK demonstrate that the non-nuclear policy of the ROK government, since first announced in 1991, has not been successfully absorbed in the minds of the general public. This is nothing less than virtual failure of the ROK non-nuclear policy in the domestic dimension, which is not surprising. Compared to advanced countries such as Japan, the ROK government has shown poor performance in managing public relations or developing education programs regarding the values of non-nuclear policy, benefits of nuclear energy and the dangers of nuclear proliferation.

Mistaken beliefs, lack of proper knowledge, and ignorance of international non-proliferation trends, etc. are bases on which excessive emotional sentiments regarding South Korean possession of nuclear weapons are resided. This domestic failure of the non-nuclear policy illustrates urgency and seriousness of public education and indicates a future direction of the ROK government's non-nuclear policy. The direction is to establish a solid system of public and school education programs to guide South Korean public—old and young—to have better understanding of dangers of nuclear weapons and importance of non-proliferation. Such an educational system will form a sound,

---

[116] *Donga Ilbo*, August 6, 2005.

solid and sustainable public basis to support the ROK government's non-nuclear policy in the future.

Considering the weight of nuclear power in the daily lives of South Korean public,[117] existing public relations campaign of the ROK government must have barely accomplished what needed to be done. Occasional displays of the benefits and safety of nuclear energy in TV commercials or newspaper advertisements are not enough and more systematic programs must be developed.

In this respect, enormous efforts and excellent performance of the Japanese government and nuclear community are highly contrasted. The Japan Foundation of Promoting Nuclear Energy Culture is largely in charge of public education. For instance, the foundation regularly—about 40 times a year—dispatches nuclear energy experts to schools without charge and delivers special education on nuclear energy. If a group of more than four people requests a lecture on nuclear energy, the foundation also sends an expert to them. Such an *ad hoc* education occurs around 300 times a year. The foundation also produces a variety of audio-visual aids on nuclear energy and distributes them to the public. It also equips itself with hundreds of videotapes on nuclear energy, radiation, climate change, nuclear fuel cycle, safety and production and rents them to the ordinary Japanese people.

Japan celebrates October 26th as an atomic energy day. Every year, a variety of educational and cultural events are held throughout Japan to advertise the positive functions and benefits of nuclear

---

[117] For instance, South Korea is the world's sixth largest nuclear power country. As of 2005, it has 17 operating nuclear power plants and 6 under construction. About 38.2% of electricity in South Korea is produced from nuclear energy.

energy. These events are organized not only by the Japanese government but also by nuclear power plants, research institutes, and civilian companies that constructed nuclear power plants. The Japanese government has strengthened school education programs as well. For example, from 2003 at all schools up to high school levels, two to three hours per year have been doled out to teach students about nuclear energy. Teachers at these schools also receive proper education about nuclear energy.

The importance of public education on WMD non-proliferation as well as nuclear energy is looming large partly because it is likely to be an international agendum. For example, at a non-proliferation conference held in Sweden in September 2000, a proposal was submitted that all new nuclear-weapon-treaties have an obligatory clause to require member nations of a treaty to educate their people about destructive powers of nuclear weapons and international efforts to curb nuclear proliferation.[118] The United Nations also has discussed how to reinforce public education on nuclear disarmament. In light of this international trend, public education must be a serious way of expanding solid and favorable public basis upon which its non-nuclear policy resides.

By disseminating necessary information for understanding a country's non-nuclear policy to the ordinary citizens, public education should be an indispensable component of improving transparency of the non-nuclear policy. Thus, the ROK government and nuclear community are advised to take

---

[118] Hiromichi Umebayashi, "Supplementary memo on a Northeast Asia nuclear weapon-free zone," a paper presented at the International Symposium on *Security and Nuclear Weapons in Northeast Asia*, Tokyo, Japan, October 3, 2000.

appropriate measures that can boost public relations campaign and increase public understanding of the benefits and importance of greater transparency of the ROK non-nuclear policy.

Chapter Six

Summary and Suggestions

The ROK and Nuclear Non-Proliferation in 2004

In August 2004, it was revealed that the ROK failed to report its nuclear activities on several occasions from the early 1980s to the IAEA. The incident drew great attention from the international non-proliferation community.

Following the South Korean submission of the initial report in August 2004, according to the additional protocol, the IAEA began a series of verification missions in South Korea. As a result of these inspections, the IAEA concluded that on a number of occasions, starting in 1982 and continuing until 2000, South Korea performed experiments and activities involving uranium conversion, uranium enrichment and plutonium separation and failed to report them to the IAEA in accordance with its obligations under its safeguard agreement.

Although the failure of reporting was observed as a matter of serious concern by the IAEA, the 2004 incident, in no way, should be interpreted as representing a desire by the ROK government to pursue a nuclear weapon development program. The IAEA acknowledged that the quantities of nuclear material involved in the experiments have not been significant and that there was no indication of further undeclared experiments.

Nevertheless, the incident has produced several important implications. An obvious conclusion is that the South Korean nuclear establishment is poorly regulated. The incident also stirred already troubled waters in Northeast Asia due to the North Korea nuclear crisis. At the same time, it was demonstrated that states will eventually pay a price if they allow nuclear research establishments to conduct activities without stringent education of the importance of adhering to non-proliferation norms and rules.

The ROK reaction was prompt and clear. The ROK government declared a new non-nuclear policy on September 18, 2004—"the four principles of peaceful uses of nuclear energy." And one of the principles was to express South Korea's determination to firmly hold on to the principle of nuclear transparency and reinforce international cooperation, including full collaboration with the IAEA inspection.

Transparency: Concept and Phenomena

Transparency began to draw the attention of academic community from the early 1990s. Universal interest of the concept of transparency has occurred in parallel with the globalization of the international community. In this respect, transparency is both a cause and an effect of lowering national borders and bureaucratic hurdles.

Transparency allows three important functions and creates compliance dynamic: (1) to permit *coordination* between actors making independent decisions, (2) to provide *reassurance* to actors cooperating or complying with the norms of the regime that they are not being taken advantage of, and (3) to exercise *deterrence* on actors contemplating non-compliance or defection.

Transparency is spreading as part and parcel of three trends— democratization, globalization and dramatic advances in technology. Increasingly, in issues ranging from security to commerce to economics, transparency is the preferred means of enforcement. In fact, the international community is embracing new standards of conduct enforced by willful disclosure.

In the realm of security, the best example is that the verification provisions in various arms control treaties have been strengthened. It is really outstanding that the arms control treaties agreed in parallel with the dissolution of the Cold War had encompassed greater transparency. For instance, the 1990 Conventional Forces in Europe I (CFE I) Treaty, the 1992 Strategic Arms Reduction Talks (START) Treaty, the 1992 UN Register of Conventional Armaments, the 1992 Open Skies Treaty, and the 1993 Chemical Weapons Convention (CWC) involve extensive amounts of information being revealed voluntarily and involuntarily.

Non-Proliferation and Transparency in the 21st Century

Cooperative security has emerged as a principal international security guideline in the 21st century, whose basic philosophy is that the appropriate principle for dealing with new security threats is cooperative engagement. Transparency is a principal element for promoting cooperative security. Specifically, it is a practical tool to embody the strategic principle of cooperative security—cooperative engagement—into reality. Three motivations are proposed that have made transparency flourish—technology development, democratization, and globalization.

It is noted that an international trend has awakened many nations of the danger of WMD and driven them to strengthen the non-proliferation regimes to prevent further proliferation. Since January 1992, in particular, the United Nations has defined that proliferation of nuclear, chemical and biological weapons as well as their means of delivery constitutes a threat to international peace and security. Thereafter, a number of important steps have been taken, on international, regional and national bases in order to curb proliferation of WMD, missiles, related materials and technologies, and terror activities.

While a series of important efforts have been undertaken in order to curb proliferation of WMD, it is noted that several events have occurred that could hamper such non-proliferation efforts. These events could kindle long-held complaints about "discrimination"—the nuclear weapon states, particularly the United States, which have treated non-nuclear weapon states with double standard, giving a favor to certain countries at Washington's own discretion.

The diffusion of civil and military technologies and the proliferation of dual-use items are a new reality faced by the international export control community since the end of the Cold War. The traditional control policy of denying access to materials and information has been undermined in this new reality. Current trends in the proliferation of dangerous technologies compel a decisive shift in the policy. One element of the policy shift should be a change in the principal mechanisms of control from denial of access to technology to cooperatively induced restraint. The key is to develop a structure that allows relatively free trade of dual-use items, while at the same time, ensuring that these items are used only in civilian applications. In this respect, the importance of enhancing transparency of supplying countries' export control policies cannot be emphasized too much.

Greater Nuclear Transparency and the ROK

Enhancing transparency of non-nuclear policy is vital for maximizing national interests of the ROK in several ways.

- The ROK will receive enormous disadvantages if it does not eliminate nuclear suspicions. An obscure non-nuclear policy could harm the credibility of South Korea's national policies as a whole. In a tightly interwoven international society, nuclear discredit would isolate Seoul, diminish its diplomatic capabilities, thus bringing about many difficulties in key issue areas.
- A transparent non-nuclear policy is important for Korean people to acquire credibility and respect as a responsible member of international community, and to maintain the esteem and dignity. In fact, enhancing transparency is a matter of national image and pride. It is a countering force that will help strip the international community of bad images and wrong perceptions of Korean people.
- It is virtually impossible for the ROK to operate a nuclear weapon program without being detected by international supervisions. Under the circumstances, an ambiguous

non-nuclear policy would only bring about political suspicions of proliferation.

- In terms of energy security, peaceful uses of nuclear power have become a critical element of South Korea's energy policy. The problem with nuclear suspicions is that it would cause visible or invisible adverse effects that stand in the way of the ROK nuclear industry's R&D activities.
- The ROK government's solid non-nuclear policy will help resolve the North Korea nuclear crisis. It will not provide North Korea with any excuse either to justify its nuclear weapon program or to delay the negotiating process. It will set a role model that must be followed by the North Korean regime, thus producing political and diplomatic pressures upon Pyongyang. It can present South Korea with a better opportunity to play a leading role in the course of resolving the nuclear crisis. This opportunity comes from both moral and practical strength based on its credible non-nuclear policy.
- Lack of transparency in the ROK non-nuclear policy and resultant nuclear suspicions will keep neighboring states in constant nervousness. This might induce unnecessary tension and an arms race in the region—a boomerang, which is obviously not what the ROK government and people would like to see.
- If a unified Korea were to go nuclear, neighboring countries understandably would make every effort to stand in the way of Korean unification. Therefore, unless South and North Korea make sure that they are non-nuclear and will remain so in the future, they cannot expect the external support and assistance that will be essential in the unification process. Of course, a peaceful

unification under South Korea's terms should have an effect of removing nuclear suspicions of the international community.

The ROK Position on Nuclear Transparency

It is essential for short-term as well as long-term interests of South Korea to invent sound positions and attitudes on enhancing transparency of its non-nuclear policy.

A first step should be to increase the understanding of policy-making and scientific communities about transparency—i.e., what transparency means, what positive roles it is to play, and what negative effects it could create, etc. Only such a thorough understanding about transparency and international trends for promoting transparency can prevent reluctance and misperception in relation to transparency. It is also a precondition for the ROK government to establish clear and reasonable goals by accepting the international standards of transparency in the nuclear field. Succinct understanding and clear goals are the basis upon which directions and means for an effective non-nuclear policy are established.

Enhanced transparency in non-proliferation and safety will help the ROK government establish a sound and solid foundation that makes stable and sustainable uses of nuclear energy feasible. International support and credibility acquired by increasing transparency will make it possible to strengthen both domestic and external bases for promoting peaceful uses of nuclear energy, which must be a shortcut for booming nuclear industry.

The policy-making and nuclear communities in the ROK should have a firm grasp on a harsh reality that enhancing policy transparency in such a sensitive area as nuclear energy is a complex issue involving multiple considerations and influencing other policy-making areas. A worldwide phenomenon of globalization has made the international system integral, member nations of the system mutually dependent, major policy issue areas linked in one way or another, and various policy-related communities within and without a nation engaged through visible and invisible networking.

Proper understanding of such complexities involved in transparency will lead the ROK government to take account of a broad spectrum of factors when an allegation is raised regarding transparency of its non-nuclear policy. Only if all relevant factors are properly paid attention to in a comprehensive way, it would be feasible for South Korea to shape a credible, durable and integral non-nuclear policy.

Transparency-related problems could be prevented or its seriousness ameliorated if the ROK government's non-nuclear policy—from its formation to implementation—addresses the psychological aspect of transparency. It would be wise for the ROK government policy-making community to avoid any remarks or behaviors that could trigger suspicions or misunderstandings from the international community. It is also important to forge and keep better relations with major nuclear-supplying countries.

Delicate diplomacy is as much important as developing technical infrastructures that creates an auspicious external environment for bolstering peaceful uses of nuclear energy and eliminates suspicions and concerns of the international community about South Korea's nuclear intention. The ROK foreign affairs community must have a firm grasp on the fact that non-proliferation is indeed one of the most important consensuses of the international community in the 21st century. With this firm acknowledgement, the ROK diplomacy in non-proliferation affairs will have to be directed toward less exclusiveness, better compliance, less secrecy and more transparency.

A domestic obstacle to promoting peaceful uses of nuclear energy in South Korea is a groundless, emotional, but widely spread national sentiment for developing nuclear weapons. Mistaken beliefs, lack of proper knowledge, and ignorance of international non-proliferation trends, etc. are bases of such excessive emotional sentiments. This illustrates urgency and seriousness of public education and indicates a future direction of the ROK non-nuclear policy. The direction is to establish a solid system of public and school education programs to guide South Korean public—old and young—to have better understanding of dangers of nuclear weapons and importance of non-proliferation. Such an educational system will form a sound, solid and sustainable public basis to support the ROK government's non-nuclear policy in the future.

Introductory Statement of the Board of Governors by IAEA Director General Dr. Mohamed ElBaradei, September 13, 2004 (Excerpts)

Implementation of the NPT Safeguards Agreement in the Republic of Korea

The Republic of Korea (ROK) brought its additional protocol into force in February 2004. Last month, in connection with the submission of the ROK's initial declarations pursuant to the additional protocol, the ROK informed the Agency that, in 2000, laboratory scale experiments involving the enrichment of uranium—using the atomic vapour laser isotope separation (AVLIS) method—had taken place at the Korea Atomic Energy Research Institute. The Agency promptly sent a team to the ROK, headed by the Director of the Safeguards Operations Division concerned, to verify this and other information.

The inspection team visited the facility where these experiments and associated physics measurements of uranium are said to have taken place, as well as other facilities where the ROK stated that its scientists had conducted uranium conversion activities in the 1980s. One of these conversion activities, which took place at three facilities that had not been declared to the Agency, involved the production of about 150 kilograms of natural uranium metal, a small amount of which, according to the ROK, was later used in the AVLIS experiments.

The ROK authorities have pointed out that the uranium enrichment experiments took place in the context of a broader experimental effort to apply AVLIS techniques to a wide range of stable isotopes. According to the ROK, only about 200 milligrams of enriched uranium were produced.

During the same trip, the inspection team visited another facility for which the results of environmental samples had revealed the presence of slightly irradiated depleted uranium with associated plutonium. The ROK authorities informed the Agency that, in the early 1980s, a laboratory scale experiment had been performed at this facility to irradiate 2.5 kilograms of depleted uranium and separate a small amount of plutonium.

The ROK authorities have stated that all the above experiments were performed without the knowledge or authorization of the ROK Government.

With the full cooperation of the ROK, the team was able, at each of the facilities visited, to examine the associated records available, perform measurements, take photographs, collect environmental samples, interview a number of the scientists involved, and view the dismantled equipment that the ROK stated had been associated with these experiments. The team was also able to place Agency seals on major components of the dismantled equipment and associated nuclear materials.

It is a matter of serious concern that the conversion and enrichment of uranium and the separation of plutonium were not reported to the Agency as required by the ROK safeguards agreement.

The Agency will continue its investigation of all aspects relevant to this new information. I will report to the Board as appropriate, and not later than at its meeting in November. I would ask the Republic of Korea to continue to provide active cooperation and maximum transparency, in order for the Agency to gain full understanding of the extent and scope of these previously undeclared activities, and to verify the correctness and completeness of ROK's declarations relevant to its nuclear programme.

Transcript of the Director General's Press Statement on IAEA Inspection in Iran, Libya & the Republic of Korea, September 13, 2004 (Excerpts)

Afternoon Statement

With regard to the Republic of Korea, I reported on the new information that came to our knowledge that there was enrichment activity at the experimental level in 2000, and that there was also some separation of plutonium in the early 80s. We obviously have sent a team to Seoul upon ROK informing us of these activities. We still have a lot of work to do. We are getting active co-operation by the Republic of Korea and I hope that co-operation will continue. I will be in a position in November to give a full written report on these activities, including its nature and scope. And hopefully be able by that time to assure the international community that these activities are isolated activities and that all measures have been taken to ensure their non-recurrence. Clearly, any activities that involve separation of plutonium or enriching of uranium are matters of serious concern from a proliferation perspective and therefore we are going to treat them with the seriousness they deserve.

Question
What about reports that enrichment in the Republic of Korea was up to 77%, very close to being bomb grade level. And secondly, do you think in Iran, uranium conversion is part of the agreement to suspend enrichment related activities?

DG's Answer
Well, on the level of enrichment in Korea, Michael, I think we, I would like to wait until we go and do our technical measurements. I know that the average enrichment in Korea was about 10%, there could be some higher peak. But I would like to wait until we do our measurements. On your second question, uranium conversion has always been a

controversial issue, whether that has been part of the suspension or not. Iran has stated on a number of occasions that they never accepted suspension with regard to the conversion. At an earlier stage, the Agency thought that they were, that it was part of the suspension, but they made it clear that they never made a commitment to have conversion as part of the suspension.

Question:
Some members of the Board have expressed the view that the South Korean issue should be reported to the Security Council. Do you share this view? And second part of the question, do you really think that this issue can be dealt with by November, given that new elements and details seem to be coming out daily?

DG's Answer
First of all we need to understand the nature and scope of the activities that took place in the Republic of Korea before we discuss what sort of action the Board needs to take. I think that the Board, at this stage, will simply ask me to continue to investigate the initial report we have received. And it will take us time, I would hope we can finish by November, but if not, then we will continue. Again, it depends on what we see; it depends of the level of co-operation we get from South Korea. But, as I said, so far, I am getting good transparency and good co-operation from Korea and I'll hope we should get a comprehensive report and get to the bottom of this issue by November. Thank you very much.

Implementation of the NPT Safeguards Agreement in the Republic of Korea, GOV/2004/84, November 26, 2004

A. Background

1. The Agreement between the Republic of Korea (ROK) and the IAEA for the Application of Safeguards in Connection with the Treaty on the Non-Proliferation of Nuclear Weapons (the Safeguards Agreement)[119] entered into force on 14 November 1975. The Additional Protocol to the Safeguards Agreement (the Additional Protocol)[120] was signed on 21 June 1999 and entered into force on 19 February 2004.

2. On 23 August 2004 the ROK, in connection with the submission of its initial declaration pursuant to the Additional Protocol, informed the Secretariat that the ROK Government had discovered, in June 2004, that laboratory scale experiments involving the enrichment of uranium using the atomic vapour laser isotope separation (AVLIS) method had been carried out, in 2000, by scientists at the Korea Atomic Energy Research Institute (KAERI) in Daejeon. The ROK explained further that the uranium enrichment experiments had taken place in the context of a broader experimental effort to apply AVLIS techniques to non-nuclear materials such as gadolinium, thallium and ytterbium. The ROK stated that only about 200 mg of enriched uranium were produced, following which the experiments were terminated, and the installation where these experiments had been carried out had been dismantled.

3. Following the ROK's submission of its initial declaration and explanation concerning the discovery of certain experiments as noted in paragraph 2 above, the Agency promptly despatched an inspection team to the ROK to verify this and other related information. From 30 August to 4 September 2004, the Agency inspection team visited the KAERI site where these experiments and associated physics

---

[119] The Safeguards Agreement is reproduced in document INFCIRC/236.
[120] The Additional Protocol is reproduced in document INFCIRC/236/Add.1.

measurements of uranium were stated by the ROK to have taken place, and also visited the manufacturers of laser components. In its investigation of the origin of the nuclear material used in the AVLIS experiments, the Agency visited: the Youngnam Chemical Plant in Ulsan; the Korea Institute of Science and Technology (KIST) in Seoul; related facilities at the KAERI site in Daejeon; and, in a follow-up verification mission, the former Goesan coal mine.

4. Subsequent Agency verification missions were carried out from 20 to 24 September and from 3 to 6 November 2004.

5. During the Agency verification missions, the ROK stated that its scientists had conducted uranium conversion activities in the 1980s which involved the production of about 154 kg of natural uranium metal, a small amount of which was later used in the AVLIS experiments.

6. The Agency inspection team also visited the TRIGA Mark III (TRIGA III) research reactor at the KAERI site in Seoul. Environmental samples, collected previously at this site, had revealed the presence of slightly irradiated depleted uranium (DU) with associated plutonium. In response to an Agency enquiry, the authorities in the ROK stated that in the early 1980s laboratory scale experiments had been performed at this facility to irradiate 2.5 kg of DU and to study the separation of uranium and plutonium. The authorities in the ROK have stated that all the above experiments were performed without the knowledge or authorization of the Government.

7. In response to an enquiry by the Agency, based on open source information, the ROK provided information on 21 October 2004 on an experiment carried out during the period from 1979 through 1981 to assess a chemical exchange process to confirm the feasibility of producing 3% U-235.

8. At the Board of Governors meeting on 13 September 2004, the Director General informed the Board that an inspection was under way, and noted that it was "a matter of serious concern that the conversion and enrichment of uranium and the separation of plutonium were not reported to the Agency as required by the ROK Safeguards

Agreement." The Director General undertook to report to the Board regarding this matter as appropriate and not later than at its meeting in November 2004, and asked the ROK "to continue to provide active cooperation and maximum transparency, in order for the Agency to gain full understanding of the extent and scope of these previously undeclared activities, and to verify the correctness and completeness of the ROK's declarations."

9. With the active cooperation of the ROK, the Agency inspection team has been able, at each of the facilities and locations visited, to examine associated records that were made available, perform measurements, take photographs, collect samples, interview a number of the scientists involved, and view the dismantled equipment that the ROK stated had been associated with the experiments. The team was also able to place Agency seals on major components of the dismantled equipment and associated nuclear material.

10. This report provides information on the nature of the safeguards issues involved, the Agency's findings and the corrective actions that have been taken by the ROK.

B. Atomic Vapour Laser Isotope Separation (AVLIS)

11. On 10 December 2002 and again on 1 April 2003, the Agency requested permission from the ROK as a transparency measure to visit KAERI's Laser Technology R&D Centre in Daejeon, in order to confirm the nature of activities undertaken at the Centre. Both requests were refused by the ROK. Following the entry into force of the ROK's Additional Protocol, the Agency was allowed to visit the Centre in March 2004, but the ROK did not permit the Agency to take environmental samples. The ROK stated that samples could be taken only after it had submitted the Article 2.a declaration under the Additional Protocol. At the same time, the ROK continued to affirm that its laser enrichment technology R&D programme did not involve the use of any nuclear material.

12. Contrary to its earlier statements, the ROK informed the Agency on 23 August 2004, in its initial declaration pursuant to its Additional Protocol, that past activities had involved laser isotope separation of

uranium. The ROK provided a summary of the experiments and the results on 23 August 2004, and informed the Agency that:

a. The ROK had enriched uranium in three separate experiments between January and February 2000 using laser isotope separation (AVLIS) technology developed by KAERI at Daejeon;

b. The amount of nuclear material used as feedstock in the enrichment experiments was 3.5 kg of natural uranium (NU) metal;

c. The AVLIS experiments had achieved an average enrichment level of 10.2% U-235 and up to 77% U-235, and had produced 200 mg of enriched uranium;

d. The laser equipment used for the uranium enrichment experiments had been dismantled, and this equipment, together with the associated material, was available for verification by the Agency; and

e. The laser enrichment activities carried out at KAERI in Daejeon had only recently come to the attention of the Government of the ROK.

Assessment of AVLIS

13. Based on the information provided by the ROK to the Agency during its recent verification missions, elementary laser research at KAERI began in the mid-1960s and continued with the development of molecular laser isotope separation (MLIS) in the 1970s and 1980s, and AVLIS technologies in the 1990s. The ROK's laser technology development involved foreign assistance. The ROK continues to develop AVLIS technologies for the separation of stable isotopes, and this programme involves the development of small, high power, solid state lasers that could be suitable for uranium enrichment. The Agency has confirmed that the declared laser equipment involved in the undeclared enrichment experiments has been dismantled and the major components of the separation system have been placed under Agency seal.[121]

---

[121] However, some of the dismantled equipment for the AVLIS experiments is being re-used by the ROK in its stable isotope separation programme (non-nuclear activities).

14. The ROK declared during the last Agency verification mission that spectroscopy work with uranium started in 1990. After reviewing information provided by the ROK, the Agency has assessed that in 1993 and 1994, the ROK carried out a uranium evaporation test involving the use of exempted DU, followed by further spectroscopy experiments during the period from 1994 to 1996 involving exempted DU and imported NU metal. The AVLIS experiments were conducted during January, February and May 2000 using indigenously produced, undeclared NU metal.

15. According to the information provided by the ROK, it appears that at least ten AVLIS related experiments involving exempted DU and undeclared NU were carried out at KAERI facilities between 1993 and 2000. The sequencing of these experiments was: uranium evaporation; spectroscopy; and uranium isotopic separation. The ROK has stated that these experiments were authorized only by the President of KAERI in Daejeon, involved some 14 KAERI scientists, and were conducted in the broader context of a stable isotope separation project. The Agency will investigate this matter further.

16. As a result of its verification activities at the KAERI site in Daejeon since August 2004, the Agency's assessment confirms the statement by the ROK that: (i) the AVLIS experiments were laboratory-scale; and (ii) the amounts of uranium involved and the enriched uranium produced were relatively small. The levels of enrichment reported by the ROK are consistent with the Agency's calculations based on computational modelling of the experimental configuration declared by the ROK. The Agency's preliminary sample results, from the product provided by the ROK, show that the average uranium enrichment level was about 10%. The Agency is continuing to assess the results of samples taken from the AVLIS equipment (i.e. the chamber and the collector plates) and the associated products.

17. The nuclear material involved in the experiments (DU and NU metal) was required to be reported by the ROK to the Agency as provided for in the Safeguards Agreement, including in particular the requirement to provide records pertaining to the experiments and all relevant nuclear material accountancy reports, including Inventory Change Reports (ICRs). The ROK was also required to declare the

facilities where the experiments were conducted, as well as to provide their design information.

18. The Agency will study further the assistance provided by foreign sources to the ROK in the development of AVLIS technology, and will continue its investigation with a view to assessing the information provided by the ROK.

C. Uranium Conversion

19. The ROK informed the Agency during its recent verification missions that it had acquired source material from two separate origins: (a) uranium ore from a former coal mine in Goesan that was processed into about 25 kg of uranium in yellowcake at a pilot milling plant at KAERI in Daejeon; and (b) uranium bearing phosphate ore imported from abroad that was processed at the Youngnam Chemical Plant in Ulsan. The ROK stated that the uranium used in the AVLIS related experiments came from the Youngnam Chemical Plant.

Assessment of Conversion Activities

20. The declaration submitted by the ROK on 23 August 2004 did not include all its conversion activities. Some of the ROK's activities involving conversion of natural UF4 to uranium metal were revealed only as a result of the Agency's verification activities.

21. The approximately 2500 kg of ammonium uranyl tricarbonate (AUT) and the approximately 100 kg of U_3O_8 recovered from uranium bearing phosphate ore, as declared by the ROK, were consistent with the records provided to the Agency. However, it is not possible for the Agency to confirm the amount of uranium that was produced either indigenously from the ore or from the imported phosphate because the ROK has dismantled the relevant plant. The Agency's results of the samples taken from the material stated by the ROK to have been indigenously produced in the former Goesan coal mine show that the material is DU rather than NU as would be expected. The ROK has provided further information on 8 November 2004, which the Agency is currently assessing.

22. During the Agency's recent verification missions, the ROK stated that it previously had three laboratories capable of producing uranium metal. Two of these laboratories were involved in the production of about 154 kg of NU metal. The third laboratory, the largest of the three, was stated by the ROK not to have been used in the production of NU metal but only for the production of DU metal. The Agency will continue to assess the total amount of the material produced in these laboratories. According to the ROK, all three laboratories were dismantled in 1994.

23. Although the records provided by the ROK are consistent with the ROK declaration, the Agency is unable to confirm the scale of NU metal production because the laboratories no longer exist. The Agency's analysis and assessment of the relatively high losses reported by the ROK in the purification and metal reduction processes are ongoing.

24. The Agency has verified the declared yellowcake and the remaining 133 kg NU metal. When the Agency has access to the dismantled conversion equipment, it will assess the capability of this equipment. In addition, the Agency is currently assessing whether the uranium recovered from phosphate ore had, upon purification to UO_2 or UF_4, a composition and purity suitable for fuel fabrication or for being isotopically enriched, before it was converted to metal.

25. The ROK was required, pursuant to its Safeguards Agreement, to report the NU converted to metal and to submit updated design information for the two facilities where the NU metal was processed. The ROK was also required to submit updated design information for the facility[122] that was used for DU metal production. The main outstanding issues regarding the ROK's previously undeclared conversion activities include the examination and assessment by the Agency of the dismantled equipment stored as waste and the presence of DU in yellowcake samples said to be originating from the former Goesan mine.

---

[122] DU metal production was undertaken in the "Uranium Ore Processing Facility."

D. Plutonium Separation

26. In November 1997, the Agency detected two particles of slightly irradiated DU with plutonium in environmental samples taken from hot cells associated with the TRIGA III reactor in Seoul. As this finding was not consistent with any declared activities by the ROK, the Agency began to investigate whether the ROK had conducted undeclared plutonium separation activities, but since at that time the routine use of environmental sampling at hot cells was a relatively new technique at the Agency, the results were treated with some caution. In December 1999, the Agency initiated consultations with the ROK, but the ROK did not acknowledge at that time having conducted any plutonium separation activity.

27. In October 2003, the results of a subsequent set of samples from the TRIGA III hot cell collected earlier confirmed the previous findings. In December 2003, the Agency requested the ROK to provide an explanation. On 31 March 2004, the ROK stated, in a letter to the Agency, that a plutonium separation experiment had been conducted at the TRIGA III hot cell. The ROK explained that, during the period from July to December 1981, a 5-pin mini fuel assembly (mini-assembly) containing about 2.5 kg of DU had been irradiated for 82 days in the TRIGA III research reactor. The laboratory-scale experiments were said to be conducted to study the separation of uranium and plutonium. The ROK elaborated that the mini-assembly had been subsequently dismantled and dissolved, between April and May 1982, as part of a basic study on the chemical characteristics of irradiated nuclear material, and that, on 30 September 1983, it reported the "test specimen" (i.e. the mini-assembly) to the Agency as a measured discard of an unirradiated assembly.

Assessment of Plutonium Separation

28. The mini-assembly fabricated at KAERI in Daejeon was transferred to the TRIGA III reactor in Seoul on 20 July 1981, at which time the Agency was notified of its transfer. The ROK submitted the required Inventory Change Report (ICR) to the Agency on 31 July 1981.

29. The ROK has stated that the mini-assembly was irradiated in the TRIGA III reactor core, and then transferred to a hot cell for heavy metal separation based on the PUREX process. After dissolution of the mini-assembly, a basic solvent extraction procedure was performed on a portion of the dissolved solution, and ion exchange used in an attempt to recover a purified plutonium product. According to the ROK declaration, "only an aqueous solution mixed with uranium, plutonium and supposedly other fission products was obtained for analysis. Quantity of the plutonium in the solution is not known," but is expected by the ROK to be less than 40 mg.

30. The plutonium separation experiment was performed in April and May 1982, contrary to the ROK's Physical Inventory Listing report, dated 31 May 1982, that the mini-assembly was still in the TRIGA III reactor core at that time. While the ROK reported to the Agency the irradiation of the mini-assembly it did not report the uranium.plutonium solution as required by the Safeguards Agreement.

31. During the recent verification missions, the ROK provided documentation regarding the irradiation history of the mini-assembly in the TRIGA III reactor. ROK officials have stated that no operating records or technical reports remain for the plutonium separation experiment.

32. In July 1984, the equipment used for the plutonium separation experiment was dismantled and, together with the product and waste solutions, transported in 1987 to KAERI in Daejeon for storage. The uranium.plutonium solution obtained in the separation experiment was not recorded by the ROK in the material accountancy records of the TRIGA III reactor nor was it reported to the Agency.

33. On 5 November 2004, the ROK stated that 0.7 g of plutonium was produced in the irradiated mini-assembly. The Agency's assessment is that the amount of plutonium produced would have been of the same order of magnitude with an isotopic content of about 98% of Pu-239.

34. The Agency has confirmed from sample analyses that the plutonium separation experiment could not have been conducted later than 1982. The Agency has assessed that although the separation equipment used

in the experiment was rudimentary, it could have been capable of recovering pure plutonium in small amounts. The dismantled equipment and the uranium.plutonium solution have been placed under Agency seals. Based on the information available, the Agency's preliminary assessment is that only one plutonium separation experiment was carried out at the KAERI site in Seoul. The ROK has stated that the experiment was conducted solely to satisfy the scientific interest of the scientists involved.

35. The plutonium separation experiment was carried out by the ROK in a safeguarded facility and was not declared to the Agency. The ROK has not provided to the Agency updated design information of the process, including the general layout of important items of equipment used in the plutonium separation experiment, as required by the ROK Safeguards Agreement. The separation experiments, the uranium-plutonium solution and the associated waste were not reported to the Agency as required by the Safeguards Agreement. Moreover, the ROK incorrectly reported the mini-assembly as a measured discard of an unirradiated fuel assembly.

36. The open issues regarding the ROK's previously undeclared plutonium separation experiment include provision by the ROK to the Agency of: relevant operating records of the plutonium separation experiment and/or detailed information about the process; and information on the results of the plutonium separation experiment and on whether any use was made of those results.

E. Chemical Enrichment Experiment

37. In response to an enquiry by the Agency based on open source information, the declaration submitted by the ROK on 21 October 2004 included information on a chemical enrichment experiment that had not been previously declared to the Agency pursuant to the Safeguards Agreement. The experiment was carried out during the period from 1979 through 1981, and was designed to assess a chemical exchange process to confirm the feasibility of producing low enriched uranium (3% U-235) for pressurized water reactor (PWR) fuel. As stated by the ROK, the experiment was carried out using 700 g of NU (UO$_2$) powder, and utilized an ion exchange column process to produce a very small

quantity of very slightly enriched uranium (0.72% U-235). The ROK stated that the project was terminated in 1981, and the equipment subsequently dismantled in 1982. During the Agency's 3.6 November verification mission, swipe samples were taken in the room where the ROK stated that the experiment was performed. During this mission the ROK also stated that the UO_2 was under safeguards; however, the use of 700 g of NU (UO_2) powder in the experiment was not reported to the Agency. The Agency is in the process of assessing the ROK's declaration regarding this matter.

F. Findings

38. On a number of occasions, starting in 1982 and continuing until 2000, the ROK conducted experiments and activities involving uranium conversion, uranium enrichment and plutonium separation, which it failed to report to the Agency in accordance with its obligations under its Safeguards Agreement. These failures are as follows:

a. Failure to report nuclear material used in evaporation, spectroscopy and enrichment experiments (AVLIS and chemical exchange) and the associated products;

b. Failure to report the production, storage and use of NU metal and associated process loss of nuclear material, and the production and transfer of waste resulting therefrom;

c. Failure to report the dissolution of an irradiated mini-assembly and the resulting uranium. plutonium solution, including the production and transfer of waste; and

d. Failure to report initial design information for the enrichment facilities and updated design information for the facilities involved in the plutonium separation experiment and the conversion to NU and DU metal.

39. The ROK has taken corrective actions by providing relevant ICRs.

40. Following the information provided by the ROK on its previously undeclared nuclear experiments, the ROK has provided active

cooperation to the Agency in providing timely information, and access to personnel and locations, and has permitted the collection of environmental and other samples for Agency analysis and assessment. The ROK should make every effort, however, to provide the operating records for the plutonium separation and uranium spectroscopy experiments and/or detailed information regarding these experiments.

41. Although the quantities of nuclear material involved have not been significant, the nature of the activities, uranium enrichment and plutonium separation, and the failures by the ROK to report these activities in a timely manner, in accordance with its obligations under its Safeguards Agreement, is (as stated by the Director General at the Board of Governors meeting on 13 September 2004) a matter of serious concern. However, based on the information provided by the ROK and the verification activities carried out by the Agency to date, there is no indication that the undeclared experiments have continued. The Agency is continuing the process of verifying the correctness and completeness of the ROK's declarations pursuant to the Safeguards Agreement and Additional Protocol.

42. The Director General will continue to report to the Board of Governors as appropriate.

IAEA Board of Governors Chairman's Conclusion on Implementation of the NPT Safeguards Agreement in the Republic of Korea, **November 26, 2004**

IAEA Board of Governors Chairman's Conclusion

The Board took note of, and expressed appreciation for, the Director General's report contained in document GOV/2004/84.

The Board shared the Director General's view that given the nature of the nuclear activities described in his report, the failure of the Republic of Korea to report these activities in accordance with its safeguards agreements is of serious concern.

At the same time, the Board noted that the quantities of nuclear material involved have not been significant, and that to date there is no indication that the undeclared experiments have continued.

The Board welcomed the corrective actions taken by the Republic of Korea, and the active cooperation it has provided to the Agency.

The Board encouraged the Republic of Korea to continue its active cooperation with the Agency, pursuant to its Safeguards Agreement and Additional Protocol.

The Board observed that the Republic of Korea has an Additional Protocol in force and that developments in the Republic of Korea demonstrate the utility of the Additional Protocol.

The Board requested that the Director General report as appropriate.

Article VII—Staff, *Statute of the International Atomic Energy Agency*, October 26, 1956

ARTICLE VII—Staff

A. The staff of the Agency shall be headed by a Director General. The Director General shall be appointed by the Board of Governors with the approval of the General Conference for a term of four years. He shall be the chief administrative officer of the Agency.

B. The Director General shall be responsible for the appointment, organization, and functioning of the staff and shall be under the authority of and subject to the control of the Board of Governors. He shall perform his duties in accordance with regulations adopted by the Board.

C. The staff shall include such qualified scientific and technical and other personnel as may be required to fulfill the objectives and functions of the Agency. The Agency shall be guided by the principle that its permanent staff shall be kept to a minimum.

D. The paramount consideration in the recruitment and employment of the staff and in the determination of the conditions of service shall be to secure employees of the highest standards of efficiency, technical competence, and integrity. Subject to this consideration, due regard shall be paid to the contributions of members to the Agency and to the importance of recruiting the staff on as wide a geographical basis as possible.

E. The terms and conditions on which the staff shall be appointed, remunerated, and dismissed shall be in accordance with regulations made by the Board of Governors, subject to the provisions of this Statute and to general rules approved by the General Conference on the recommendation of the Board.

F. In the performance of their duties, the Director General and the staff shall not seek or receive instructions from any source external to the Agency. They shall refrain from any action which might reflect on their position as officials of the Agency; subject to their responsibilities to the Agency, they shall not disclose any industrial secret or other confidential information coming to their knowledge by reason of their official duties for the Agency. Each member undertakes to respect the international character of the responsibilities of the Director General and the staff and shall not seek to influence them in the discharge of their duties.

G. In this article the term "staff" includes guards.

United Nations Security Council Resolution 1540, April 28, 2004

The Security Council,

Affirming that proliferation of nuclear, chemical and biological weapons, as well as their means of delivery,[*] constitutes a threat to international peace and security,

Reaffirming, in this context, the Statement of its President adopted at the Council's meeting at the level of Heads of State and Government on 31 January 1992 (S/23500), including the need for all Member States to fulfill their obligations in relation to arms control and disarmament and to prevent proliferation in all its aspects of all weapons of mass destruction,

Recalling also that the Statement underlined the need for all Member States to resolve peacefully in accordance with the Charter any problems in that context threatening or disrupting the maintenance of regional and global stability,

Affirming its resolve to take appropriate and effective actions against any threat to international peace and security caused by the proliferation of nuclear, chemical and biological weapons and their means of delivery, in conformity with its primary responsibilities, as

---

[*] Definitions for the purpose of this resolution only: *Means of delivery*: missiles, rockets and other unmanned systems capable of delivering nuclear, chemical, or biological weapons, that are specially designed for such use. *Non-State actor*: individual or entity, not acting under the lawful authority of any State in conducting activities which come within the scope of this resolution. *Related materials*: materials, equipment and technology covered by relevant multilateral treaties and arrangements, or included on national control lists, which could be used for the design, development, production or use of nuclear, chemical and biological weapons and their means of delivery.

provided for in the United Nations Charter,

Affirming its support for the multilateral treaties whose aim is to eliminate or prevent the proliferation of nuclear, chemical or biological weapons and the importance for all States parties to these treaties to implement them fully in order to promote international stability,

Welcoming efforts in this context by multilateral arrangements which contribute to non-proliferation,

Affirming that prevention of proliferation of nuclear, chemical and biological weapons should not hamper international cooperation in materials, equipment and technology for peaceful purposes while goals of peaceful utilization should not be used as a cover for proliferation,

Gravely concerned by the threat of terrorism and the risk that non-State actors* such as those identified in the United Nations list established and maintained by the Committee established under Security Council Resolution 1267 and those to whom resolution 1373 applies, may acquire, develop, traffic in or use nuclear, chemical and biological weapons and their means of delivery,

Gravely concerned by the threat of illicit trafficking in nuclear, chemical, or biological weapons and their means of delivery, and related materials,* which adds a new dimension to the issue of proliferation of such weapons and also poses a threat to international peace and security,

Recognizing the need to enhance coordination of efforts on national, subregional, regional and international levels in order to strengthen a global response to this serious challenge and threat to international security,

Recognizing that most States have undertaken binding legal obligations under treaties to which they are parties, or have made other commitments aimed at preventing the proliferation of nuclear, chemical or biological weapons, and have taken effective measures to account for, secure and physically protect sensitive materials, such as those required by the Convention on the Physical Protection of Nuclear Materials and those recommended by the IAEA Code of Conduct on the Safety and Security of Radioactive Sources,

Recognizing further the urgent need for all States to take additional effective measures to prevent the proliferation of nuclear, chemical or biological weapons and their means of delivery,

Encouraging all Member States to implement fully the disarmament treaties and agreements to which they are party,

Reaffirming the need to combat by all means, in accordance with the Charter of the United Nations, threats to international peace and security caused by terrorist acts,

Determined to facilitate henceforth an effective response to global threats in the area of non-proliferation,

Acting under Chapter VII of the Charter of the United Nations,

1. *Decides that* all States shall refrain from providing any form of support to non-State actors that attempt to develop, acquire, manufacture, possess, transport, transfer or use nuclear, chemical or biological weapons and their means of delivery;

2. *Decides also* that all States, in accordance with their national procedures, shall adopt and enforce appropriate effective laws which prohibit any non-State actor to manufacture, acquire, possess, develop, transport, transfer or use nuclear, chemical or biological weapons and their means of delivery, in particular for terrorist purposes, as well as attempts to engage in any of the foregoing activities, participate in them as an accomplice, assist or finance them;

3. *Decides also* that all States shall take and enforce effective measures to establish domestic controls to prevent the proliferation of nuclear, chemical, or biological weapons and their means of delivery, including by establishing appropriate controls over related materials and to this end shall:
   (a) Develop and maintain appropriate effective measures to

account for and secure such items in production, use, storage or transport;

(b) Develop and maintain appropriate effective physical protection measures;

(c) Develop and maintain appropriate effective border controls and law enforcement efforts to detect, deter, prevent and combat, including through international cooperation when necessary, the illicit trafficking and brokering in such items in accordance with their national legal authorities and legislation and consistent with international law;

(d) Establish, develop, review and maintain appropriate effective national export and transshipment controls over such items, including appropriate laws and regulations to control export, transit, transshipment and re-export and controls on providing funds and services related to such export and transshipment such as financing, and transporting that would contribute to proliferation, as well as establishing end-user controls; and establishing and enforcing appropriate criminal or civil penalties for violations of such export control laws and regulations;

4. *Decides* to establish, in accordance with rule 28 of its provisional rules of procedure, for a period of no longer than two years, a Committee of the Security Council, consisting of all members of the Council, which will, calling as appropriate on other expertise, report to the Security Council for its examination, on the implementation of this resolution, and to this end calls upon States to present a first report no later than six months from the adoption of this resolution to the Committee on steps they have taken or intend to take to implement this resolution;

5. *Decides* that none of the obligations set forth in this resolution shall be interpreted so as to conflict with or alter the rights and obligations of State Parties to the Nuclear Non-Proliferation Treaty, the Chemical Weapons Convention and the Biological and Toxin Weapons Convention or alter the responsibilities of the International Atomic Energy Agency or the Organization

for the Prohibition of Chemical Weapons;

6. *Recognizes* the utility in implementing this resolution of effective national control lists and calls upon all Member States, when necessary, to pursue at the earliest opportunity the development of such lists;

7. *Recognizes* that some States may require assistance in implementing the provisions of this resolution within their territories and invites States in a position to do so to offer assistance as appropriate in response to specific requests to the States lacking the legal and regulatory infrastructure, implementation experience and/or resources for fulfilling the above provisions;

8. *Calls upon* all States:
   (a) To promote the universal adoption and full implementation, and, where necessary, strengthening of multilateral treaties to which they are parties, whose aim is to prevent the proliferation of nuclear, biological or chemical weapons;
   (b) To adopt national rules and regulations, where it has not yet been done, to ensure compliance with their commitments under the key multilateral non-proliferation treaties;
   (c) To renew and fulfill their commitment to multilateral cooperation, in particular within the framework of the International Atomic Energy Agency, the Organization for the Prohibition of Chemical Weapons and the Biological and Toxin Weapons Convention, as important means of pursuing and achieving their common objectives in the area of non-proliferation and of promoting international cooperation for peaceful purposes;
   (d) To develop appropriate ways to work with and inform industry and the public regarding their obligations under such laws;

9. *Calls upon* all States to promote dialogue and cooperation on non-proliferation so as to address the threat posed by

proliferation of nuclear, chemical, or biological weapons, and their means of delivery;

10. Further to counter that threat, *calls upon* all States, in accordance with their national legal authorities and legislation and consistent with international law, to take cooperative action to prevent illicit trafficking in nuclear, chemical or biological weapons, their means of delivery, and related materials;

11. *Expresses* its intention to monitor closely the implementation of this resolution and, at the appropriate level, to take further decisions which may be required to this end;

12. *Decides* to remain seized of the matter.

Statement of the Security Council at the Level of Heads of States and Government, United Nations Security Council, S/23500, January 31, 1992 (Excerpts)

Disarmament, Arms Control and Weapons of Mass Destruction

The members of the Council, while fully conscious of the responsibilities of other organs of the United Nations in the fields of disarmament, arms control and non-proliferation, reaffirm the crucial contribution which progress in these areas can make to the maintenance of international peace and security. They express their commitment to take concrete steps to enhance the effectiveness of the United Nations in these areas.

The members of the Council underline the need for all Member States to fulfil their obligations in relation to arms control and disarmament; to prevent the proliferation in all its aspects of all weapons of mass destruction; to avoid excessive and destabilizing accumulations and transfers of arms; and to resolve peacefully in accordance with the Charter any problems concerning these matters threatening or disrupting the maintenance of regional and global stability. They emphasize the importance of the early ratification and implementation by the States concerned of all international and regional arms control arrangements, especially the START and CFE Treaties.

The proliferation of all weapons of mass destruction constitutes a threat to international peace and security. The members of the Council commit themselves to working to prevent the spread of technology related to the research for or production of such weapons and to take appropriate action to that end.

On nuclear proliferation, they note the importance of the decision of many countries to adhere to the Non-Proliferation Treaty and emphasize the integral role in the implementation of that Treaty of fully effective IAEA safeguards, as well as the importance of effective export controls. The members of the Council will take appropriate measures in the case of any violations notified to them by the IAEA.

On chemical weapons, they support the efforts of the Geneva Conference with a view to reaching agreement on the conclusion, by the end of 1992, of a universal convention, including a verification regime, to prohibit chemical weapons.

On conventional armaments, they note the General Assembly's vote in favour of a United Nations register of arms transfers as a first step, and in this connection recognize the importance of all States providing all the information called for in the General Assembly's resolution.

In conclusion, the members of the Security Council affirm their determination to build on the initiative of their meeting in order to secure positive advances in promoting International peace and security. They agree that the United Nations Secretary-General has a crucial role to play. The members of the Council express their deep appreciation to the outgoing Secretary-General, His Excellency Mr. Javier Perez de Cuellar, for his outstanding contribution to the work of the United Nations, culminating in the signature of the El Salvador peace agreement They welcome the new Secretary-General, His Excellency Dr. Boutros Boutros-Ghali, and note with satisfaction his intention to strengthen and improve the functioning of the United Nations. They pledge their full support to him, and undertake to work closely with him and his staff in fulfillment of their shared objectives, including a more efficient and effective United Nations system.

The members of the Council agree that the world now has the best chance of achieving international peace and security since the foundation of the United Nations. They undertake to work in close cooperation with other United Nations Member States in their own efforts to achieve this, as well as to address urgently all the other problems, in particular those of economic and social development, requiring the collective response of the international community. They recognize that peace and prosperity are indivisible and that lasting peace and stability require effective international cooperation for the eradication of poverty and the promotion of a better life for all in larger freedom.

Joint Statement by the European Union and United States on the Joint Program of Work on the Non-proliferation of Weapons of Mass Destruction, June 20, 2005

Proliferation of weapons of mass destruction (WMD) and their delivery systems continue to be a preeminent threat to international peace and security. This global challenge needs to be tackled individually and collectively, and requires an effective global response. We are fully committed to support in that respect the important role of the United Nations Security Council and other key UN institutions.

The United States and the European Union are steadfast partners in the fight against the proliferation of weapons of mass destruction, and will undertake several new initiatives to strengthen our cooperation and coordination in this important arena, even as we enhance our ongoing efforts.

Building Global Support for Non-proliferation

The European Union and the United States will enhance information sharing, discuss assessments of proliferation risks, and work together to broaden global support for and participation in non-proliferation endeavors. We will increase transparency about our non-proliferation dialogues with other countries to ensure, to the extent possible consistency in our non-proliferation messages.

We reaffirm our willingness to work together to implement and strengthen key arms control, disarmament and non-proliferation treaties, agreements and commitments that ban the proliferation of WMD and their delivery systems. In particular we underline the importance of the Treaty on the Non-Proliferation of Nuclear Weapons (NPT), the Biological and Toxin Weapons Convention and the Chemical Weapons Convention. We will increase our effort to promote, individually or, where appropriate, jointly, the universalisation of these Treaties and

Conventions and the adherence to the Hague Code of Conduct against the proliferation of ballistic missiles.

Reinforcing the NPT

The EU and the US reaffirm that the NPT is central to preventing the spread of nuclear weapons. The EU and the US stress the urgency to maintain the authority and the integrity of the Treaty. To that end, the EU and the U.S. recommit to fulfill our obligations under the Treaty while working together in order to strengthen it. We will evaluate lessons learned from the 2005 Review Conference and continue to stress the importance of compliance with and universal adherence to the NPT.

Recognizing the Importance of the Biological Threat

The EU and the US will work together in advance of the upcoming BTWC- Review Conference in 2006, in order to strengthen the Biological Weapons and Toxin Weapons Convention.

Promoting Full Implementation of UNSCR 1540

We will coordinate efforts to assist and enhance the work being done by the UNSCR 1540 Committee, and compliance with the resolution. We will work together to respond, where possible, to assistance requests from States seeking to implement the requirements set by the UNSC Resolution 1540 and in particular, to put in place national legal regulatory, and enforcement measures against proliferation.

Establishing a Dialogue on Compliance and Verification

The European Union and the United States renew their call on all States to comply with their arms control, disarmament and non-proliferation agreements and commitments. We will seek to ensure, through regular exchanges, strict implementation of compliance with these agreements and commitments. We will continue to support the multilateral institutions charged with verifying activities under relevant treaties and agreements. We will ask our experts to discuss issues of compliance and verification in order to identify areas of possible cooperation and joint undertaking.

Strengthening the IAEA

The U.S. and the EU welcome the steps taken earlier this month by the Board of Governors of the IAEA that created a new Committee on Safeguards and Verification, which will enhance the IAEA's effectiveness and strengthen its ability to ensure that nations comply with their NPT safeguards obligations. We will work together to ensure all States conclude a comprehensive safeguards agreement and an Additional Protocol with the IAEA. We agree that the Additional Protocol should become a standard for nuclear cooperation and non-proliferation.

Advancing the Proliferation Security Initiative

As we enhance our own capabilities, laws and regulations to improve our readiness for interdiction actions, we pledge to strengthen the Proliferation Security Initiative and encourage PSI countries to support enhanced cooperation against proliferation networks, including tracking and halting financial transactions related to proliferation.

Global Partnership

The U.S. and the EU reaffirm our commitment to the Global Partnership Initiative Against the Spread of Weapons and Materials of Mass Destruction. We will assess ongoing and emerging threats and coordinate our non-proliferation cooperation, including with other participating states, to focus resources on priority concerns and to make the most effective use of our resources.

Enhancing Nuclear Security

We intend to expand and deepen cooperation to enhance the security of nuclear and radiological materials. We welcome the establishment of the Global Threat Reduction Initiative (GTRI) and will cooperate closely to implement this important new initiative, including by exploring opportunities under the GTRI to reduce the threat posed by radiological dispersal devices and by identifying specific radiological threat reduction projects that could be implemented.

Ensuring Radioactive Source Security

We remain concerned by the risks posed by the potential use of radioactive sources for terrorist purposes. We will work towards having effective controls applied by the end of 2005 in accordance with the IAEA Guidance on the Import and Export of Radioactive Sources. We will support IAEA efforts to assist countries that need such assistance to establish effective and sustainable controls.

Rationalizing Multilateral Disarmament Work

We will continue to cooperate in order to overcome the stalemate in the Conference on Disarmament and pursue reforming of the UN General Assembly's First Committee on disarmament and international security. These initiatives are part of our broader efforts to streamline and make the multilateral disarmament, arms control and non-proliferation machinery more responsive.

The U.S. and the EU take special note of the Conference to Consider and Adopt Amendments to the Convention on the Physical Protection of Nuclear Material (CPPNM) that will take place at the IAEA, July 4-8 2005, and we urge all States Parties to the CPPNM to attend and fully support adoption of an amended Convention.

International Convention for the Suppression of Acts of Nuclear Terrorism, April 13, 2005 (Excerpts)

Article 1

For the purposes of this Convention:

1. "Radioactive material" means nuclear material and other radioactive substances which contain nuclides which undergo spontaneous disintegration (a process accompanied by emission of one or more types of ionizing radiation, such as alpha-, beta-, neutron particles and gamma rays) and which may, owing to their radiological or fissile properties, cause death, serious bodily injury or substantial damage to property or to the environment.

2. "Nuclear material" means plutonium, except that with isotopic concentration exceeding 80 per cent in plutonium-238; uranium-233; uranium enriched in the isotope 235 or 233; uranium containing the mixture of isotopes as occurring in nature other than in the form of ore or ore residue; or any material containing one or more of the foregoing;

Whereby "uranium enriched in the isotope 235 or 233" means uranium containing the isotope 235 or 233 or both in an amount such that the abundance ratio of the sum of these isotopes to the isotope 238 is greater than the ratio of the isotope 235 to the isotope 238 occurring in nature.

3. "Nuclear facility" means:
   (a) Any nuclear reactor, including reactors installed on vessels, vehicles, aircraft or space objects for use as an energy source in order to propel such vessels, vehicles, aircraft or space objects or for any other purpose;
   (b) Any plant or conveyance being used for the production, storage, processing or transport of radioactive material.

4. "Device" means:
   (a) Any nuclear explosive device; or
   (b) Any radioactive material dispersal or radiation-emitting device which may, owing to its radiological properties, cause death, serious bodily injury or substantial damage to property or to the environment.

5. "State or government facility" includes any permanent or temporary facility or conveyance that is used or occupied by representatives of a State, members of a Government, the legislature or the judiciary or by officials or employees of a State or any other public authority or entity or by employees or officials of an intergovernmental organization in connection with their official duties.

6. "Military forces of a State" means the armed forces of a State which are organized, trained and equipped under its internal law for the primary purpose of national defence or security and persons acting in support of those armed forces who are under their formal command, control and responsibility.

Article 2

1. Any person commits an offence within the meaning of this Convention if that person unlawfully and intentionally:
   (a) Possesses radioactive material or makes or possesses a device:
      (i)   With the intent to cause death or serious bodily injury; or
      (ii) With the intent to cause substantial damage to property or to the environment;
   (b) Uses in any way radioactive material or a device, or uses or damages a nuclear facility in a manner which releases or risks the release of radioactive material:
      (i)   With the intent to cause death or serious bodily injury; or
      (ii) With the intent to cause substantial damage to property or to the environment; or
      (iii) With the intent to compel a natural or legal person, an international organization or a State to do or refrain from doing an act.

2. Any person also commits an offence if that person:

(a) Threatens, under circumstances which indicate the credibility of the threat, to commit an offence as set forth in paragraph 1 (b) of the present article; or

(b) Demands unlawfully and intentionally radioactive material, a device or a nuclear facility by threat, under circumstances which indicate the credibility of the threat, or by use of force.

3. Any person also commits an offence if that person attempts to commit an offence as set forth in paragraph 1 of the present article.

4. Any person also commits an offence if that person:
   (a) Participates as an accomplice in an offence as set forth in paragraph 1, 2 or 3 of the present article; or
   (b) Organizes or directs others to commit an offence as set forth in paragraph 1, 2 or 3 of the present article; or
   (c) In any other way contributes to the commission of one or more offences as set forth in paragraph 1, 2 or 3 of the present article by a group of persons acting with a common purpose; such contribution shall be intentional and either be made with the aim of furthering the general criminal activity or purpose of the group or be made in the knowledge of the intention of the group to commit the offence or offences concerned.

Article 5

Each State Party shall adopt such measures as may be necessary:
   (a) To establish as criminal offences under its national law the offences set forth in article 2;
   (b) To make those offences punishable by appropriate penalties which take into account the grave nature of these offences.

Article 6

Each State Party shall adopt such measures as may be necessary, including, where appropriate, domestic legislation, to ensure that criminal acts within the scope of this Convention, in particular where they are intended or calculated to provoke a state of terror in the general public or in a group of persons or particular persons, are under no circumstances justifiable by considerations of a political, philosophical, ideological, racial, ethnic, religious or other similar nature and are punished by penalties consistent with their grave nature.

Article 7

1. States Parties shall cooperate by:
   (a) Taking all practicable measures, including, if necessary, adapting their national law, to prevent and counter preparations in their respective territories for the commission within or outside their territories of the offences set forth in article 2, including measures to prohibit in their territories illegal activities of persons, groups and organizations that encourage, instigate, organize, knowingly finance or knowingly provide technical assistance or information or engage in the perpetration of those offences;
   (b) Exchanging accurate and verified information in accordance with their national law and in the manner and subject to the conditions specified herein, and coordinating administrative and other measures taken as appropriate to detect, prevent, suppress and investigate the offences set forth in article 2 and also in order to institute criminal proceedings against persons alleged to have committed those crimes. In particular, a State Party shall take appropriate measures in order to inform without delay the other States referred to in article 9 in respect of the commission of the offences set forth in article 2 as well as preparations to commit such offences about which it has learned, and also to inform, where appropriate, international organizations.

2. States Parties shall take appropriate measures consistent with their national law to protect the confidentiality of any information which they receive in confidence by virtue of the provisions of this Convention from another State Party or through participation in an activity carried out for the implementation of this Convention. If States Parties provide information to international organizations in confidence, steps shall be taken to ensure that the confidentiality of such information is protected.

3. States Parties shall not be required by this Convention to provide any information which they are not permitted to communicate pursuant to national law or which would jeopardize the security of the State

concerned or the physical protection of nuclear material.

4. States Parties shall inform the Secretary-General of the United Nations of their competent authorities and liaison points responsible for sending and receiving the information referred to in the present article. The Secretary-General of the United Nations shall communicate such information regarding competent authorities and liaison points to all States Parties and the International Atomic Energy Agency. Such authorities and liaison points must be accessible on a continuous basis.

Article 8

For purposes of preventing offences under this Convention, States Parties shall make every effort to adopt appropriate measures to ensure the protection of radioactive material, taking into account relevant recommendations and functions of the International Atomic Energy Agency.

References

Books

Blacker, Coit and Gloria Duffy, *International Arms Control: Issues and Agreement* (Stanford, CA: Stanford University Press, 1984).

Carter, Ashton, William Perry and John Steinbruner, *A New Concept of Cooperative Security* (Washington, D.C.: The Brookings Institution, 1992).

Chayes, Antonia and Abram Chayes, *The New Sovereignty: Compliance with International Regulatory Agreements* (Cambridge, Massachusetts: Harvard University Press, 1995).

Cheon, Seongwhun, *Non-Nuclear Policy of the Unified Korea: Looking Beyond and Being in the Process of Unification* (Seoul: KINU, 2002) (in Korean).

Levin, Norman, *The Shape of Korea's Future: South Korean Attitudes Toward Unification and Long-Term Security Issues* (Santa Monica: RAND, 1999).

Reinicke, Wolfgang, *Global Public Policy: Governing without Government?* (Washington, D.C.: The Brookings Institution, 1998).

Articles

Adler, Emanuel and Peter Haas, "Conclusion: epistemic communities, world order, and the creation of a reflective research program," *International Organization*, Winter 1992.

Blumenthal, Dan, "Facing a nuclear North Korea," *Asian Outlook*, June-July 2005, http://www.aei.org/asia.

Brin, David, "Letters: transparency's virtues," *Foreign Policy*,

Fall 1998.

Chayes, Antonia and Abram Chayes, "Regime architecture: elements and principles," in Janne Nolan, ed., *Global Engagement: Cooperation and Security in the 21ˢᵗ Century* (Washington, D.C.: The Brookings Institution, 1994).

Finkelstein, Neal, "Introduction: transparency in public policy," in Neal Finkelstein, ed., *Transparency in Public Policy: Great Britain and the United States* (London: Macmillan Press, 2000).

Florini, Ann, "The end of secrecy," *Foreign Policy*, Summer 1998.

_____, "A new role for transparency," *Contemporary Security Policy*, August 1997.

Gallagher, Nancy, "The politics of verification: why 'how much?' is not enough," *Contemporary Security Policy*, August 1997.

Glosserman, Brad, "The current situation in the field of nuclear disarmament and non-proliferation," a paper presented at the International Symposium on *Peace and Environmental Issues*, held in Kanazawa city, Japan, June 13-14, 2005.

Haas, Peter, "Introduction: epistemic communities and international policy coordination," *International Organization*, Winter 1992.

Kamp, Karl-Heinz, "Germany and the future of nuclear weapons in Europe," *Security Dialogue*, Vol. 26, No. 3, 1995.

Kang, Jungmin, *et al.*, "South Korea's nuclear surprise," *Bulletin of the Atomic Scientists*, January/February 2005.

Kwon, Hee-seog, Director of Non-proliferation and Disarmament, Ministry of Foreign Affairs and Disarmament, "Lessons and perspective of nuclear transparency in Korea," a paper presented at the Seminar on *Nuclear Energy Non-proliferation in East Asia*, organized by Korean Nuclear Society and Sandia National Laboratories, on August 24-26,

2005, in Seoul, South Korea.

Krasner, Stephen, "Structural causes and regime consequences: regimes as intervening variables," in Stephen Krasner, ed., *International Regimes* (Ithaca: Cornell University Press, 1983).

Menon, Rajan and S. Enders Wimbush, "Asia in the 21st century: power politics alive and well," *The National Interest*, Spring 2000.

Mitchell, Ronald, "Sources of transparency: information systems in international regimes," *International Studies Quarterly*, Vol. 42, 1998.

Mochiji, Toshiro, *et al.*, "Joint DOE-PNC research on the use of transparency in support of nuclear non-proliferation," *Journal of Nuclear Materials Management*, Fall 1999.

Nakane, Takeshi, "Japans' efforts in disarmament and non-proliferation after the 2005 NPT Review Conference," a paper presented at the International Symposium on *Peace and Environmental Issues*, held in Kanazawa city, Japan, June 13-14, 2005.

Nolan, Janne, "The concept of cooperative security," in Janne Nolan, ed., *Global Engagement: Cooperation and Security in the 21st Century* (Washington, D.C.: The Brookings Institution, 1994).

_____, *et al.*, "The imperatives for cooperation," in Janne Nolan, ed., *Global Engagement: Cooperation and Security in the 21st Century* (Washington, D.C.: The Brookings Institution, 1994).

Pilat, Joseph, "Arms control, verification and transparency," in Jeffrey Larsen and Gregory Rattray, eds., *Arms Control Toward the 21st Century* (Boulder: Lynne Rienner, 1996).

Reinicke, Wolfgang, "Cooperative security and the political economy of non-proliferation," in Janne Nolan, ed., *Global*

Engagement: Cooperation and Security in the 21st Century (Washington, D.C.: The Brookings Institution, 1994).

Umebayashi, Hiromichi, "Supplementary memo on a Northeast Asia nuclear weapon-free zone," a paper presented at the International Symposium on *Security and Nuclear Weapons in Northeast Asia*, Tokyo, Japan, October 3, 2000.

U.S. Department of Energy, "Policy forum: energy futures," *Washington Quarterly*, Autumn 1996.

van der Meer, Klaas, "The radiological threat: verification at the source," *Verification Yearbook 2003* (London: The VERTIC, 2003).

Documents and Periodicals

Boutros-Ghali, Boutros, *New Dimensions of Arms Regulation and Disarmament in the Post-Cold War Era*, Report of the Secretary-General, United Nations A/C 1/47/7, October 27, 1992.

The Bush Administration's Non-proliferation Policy: Successes and Future Challenges, Testimony by Under Secretary of State for Arms Control and International Security John Bolton to the House International Relations Committee, March 30, 2004.

Cooperative Threat Reduction (Washington, D.C.: U.S. Department of Defense, April 1995).

The Defense White Paper (Seoul: The Ministry of National Defense, 2004).

Gorwitz, Mark, *The South Korean Laser Isotope Separation Experience* (Washington, D.C.: Institute for Science and International Security, 2004).

Hecker, Siegfried, Senate Committee on Foreign Relations Hearings on Visit to the Yongbyon Nuclear Scientific

Research Center in North Korea, January 21, 2004.

Joint Statement Between President George W. Bush and Prime Minister Manmohan Singh, Office of the Press Secretary, the White House, July 18, 2005, http://www.whitehouse.gov/news/releases/2005/07/200507186.html.

Kerr, Paul, "IAEA continues investigation into South Korean nuclear activities," September 17, 2004, http://www.Arms control.org.

McFate, Patricia, *et al.*, *The Converging Roles of Arms Control Verification, Confidence-Building Measures, and Peace Operations: Opportunities for Harmonization and Synergies*, Arms Control Verification Studies No. 6 (Ottawa: Department of Foreign Affairs and International Trade, Canada, 1994).

Nuclear Status Report: Nuclear Weapons, Fissile Material, and Export Controls in the Former Soviet Union (Monterey: Monterey Institute of International Studies, June 2001).

Testimony of John Bolton, Under Secretary for Arms Control and International Security, U.S. Department of State, Committee on International Relations, United States House of Representatives, June 4, 2003.

The United States Commission on National Security/21st Century, *Seeking A National Strategy: A Concert For Preserving Security And Promoting Freedom*, April 15, 2000.

http://www.bellona.no/imaker?id=20093&sub=1.

http://www.bxa.doc.gov/ComplianceAndEnforcement/TECISydney7_03Principles.htm.

http://www.clw.org/control/sort/treatytext.html.

http://www.globalsecurity.org/wmd/library/policy/dod/npr.htm.

http://www.ifins.org/pages/kison-archive-kn545.thm.

http://www.nautilus.org/napsnet/dr/index.html#item2.

http://www.stratfor.com.

Asahi Shimbun.

Donga Ilbo.
Financial Times.
Korean Central News Agency.
Los Angeles Times.
New York Times.
New Yorker.
NNCA Newsletter.
Reuters.
Rodong Shinmun.
Washington Post.
Yonhap News.